WORKBOOK ON STATISTICS

FIRST EDITION

PURPOSE:

For the student

A comprehensive study aid, with examples similar to those in the examination papers of several Boards of Examiners for accountancy. Explicit worked solutions are presented together with detailed explanations of the main points in these solutions.

COVERAGE:

This manual is designed to cover all topics on statistical methods and quantitative techniques to be found in the appropriate examination papers of several accountancy bodies. Most of the examples presented are typical of examination questions and some examples have been specifically constructed in order to complete the logical development of the subject and to fill various gaps. As well as being an aid to answering examination questions, this text is intended as a powerful teaching aid for the student. The explanatory notes contain detailed discussions of the motivation and concepts behind the techniques used and a careful reading of these notes will be rewarded with a firm grasp of the subject.

THE AUTHOR:

Barry Blight MA(Cantab), Dip.Math.Stat(Cantab), MPhil(Lond), PhD(Lond), is a lecturer at Birkbeck College, University of London, and a statistical consultant; he has advised and lectured to many industrial and commercial companies.

ISBN: 0 85459 234-2

COMPANION PUBLICATIONS   Taxwise Taxation Workbook No. 1 covering income tax, corporation tax and capital gains tax.

Taxwise Taxation Workbook No. 2 covering capital transfer tax, taxation of trusts, value added tax and tax planning and tax management.

Printed and bound in Great Britain by
Whitstable Litho Ltd
Kent

CONTENTS

No.	Main Coverage	Estimated Students Working Time

GLOSSARY OF TERMS

BASIS	The set of non-zero variables in the solution to a linear programming problem.
BINOMIAL NORMAL POISSON CHI-SQUARED (χ^2) t	) Probability distributions that are relevant to drawing) conclusions from statistical data. Each of these distributions is) tabulated.))
CONTINGENCY TABLE	A two-way table showing the frequencies arising when items (or people) are classified according to two attributes. The subsequent analysis uses these observed frequencies to investigate the association between these attributes.
DIE	The singular of dice.
DUAL PROBLEM	A secondary linear programming problem that arises from the main problem by looking at the constraints in a different way.
FLOATS (Free, independent, total)	Time measurements associated with a branch of a network representing the amount of slack there is for the job represented by the branch. If the job is on the critical path the floats will be zero.
HISTOGRAM	The frequency distribution of a data set. This is sometimes called a bar chart.
INDEX NUMBERS	The rescaling of a time series for ease of comparison and assessment.
MEAN, MEDIAN, MODE	Statistical measurements of location.
OGIVE	A diagram of the cumulative frequencies of a data set.
QUARTILE SEMI-QUARTILE RANGE STANDARD DEVIATION STANDARD ERROR VARIANCE	) Statistical measures of spread and variability.))))
SIMPLEX METHOD	A technique used to solve a linear programming problem; ideal for computer programming.
VENN DIAGRAM	A useful diagram for the interpretation and assessment of probabilities and frequencies.

DESCRIPTION OF SYMBOLS

Symbol used in text	Description	Other symbols used elsewhere
$p(\ )$ $\text{prob}(\ )$	probability of ...	$P(\ )$
p	population proportion binomial parameter	π
n	sample size	
$^{n}C_{r}$	combinatorial coefficient (in binomial distribution)	$\binom{n}{r}$
m	Poisson mean Poisson parameter	λ
Z	Standard normal variate	U, X
μ	population mean	
σ	population standard deviation	S
$\bar{x}$	sample mean	m
t	Student's t, a sampling distribution	
H_{0}, H_{1}	hypotheses (null and alternative)	
χ^{2}	chi-squared, a sampling distribution	
r	correlation coefficient	ρ
b, B	slope of regression line	β
a, A	intercept of regression line	α

1 PROBABILITY - FUNDAMENTAL LAWS AND CONCEPTS

1.1   (a)   Define the following terms as used in the study of probability and illustrate each definition with a simple example:

      (i) relative frequency

     (ii) independent events

    (iii) mutually exclusive events

    Can two exclusive events be independent?

  (b)   What is the probability of throwing at least one 6 in two throws of a die?

  (c)   Two dice are rolled; each shows a 6.

    What is the probability that, if the dice were rolled again, the same result would appear?

    Would the probability of a double 6 be different if the first roll had produced no 6's? Give reasons.

  (d)   The successful operation of three separate switches is needed to control a machine.

    If the probability of failure of each switch is 0.1, what is the probability that the machine may break down?

1.2   (a)   (i)   Let us suppose that in a large number of similar independent trials, N, an event of interest is observed to occur n times. Then the ratio n/N is known as the <u>Relative Frequency</u> of the event. As N becomes very large this 'settles down' to a value which we call the probability of the event.

As an example consider a series of tosses of a fair coin. If heads is the event of interest, then the relative frequency is the ratio of the number of heads observed to the total number of tosses. As the number of tosses becomes very large this tends to $\frac{1}{2}$.

(ii)   Two events are <u>independent</u> if the occurrence or non-occurrence of one of the events has no influence on the probability of the other.

As an example consider the simultaneous throwing of a die and tossing of a coin. Whatever the outcome of the coin toss the probability of a 6 on the die is unaffected.

(iii)   A set of events is <u>mutually exclusive</u> if no pair of the events can occur at the same time.

As an example consider the single throw of a fair die. This can result in one of 6 events corresponding to the 6 faces of the die. But only one of these events can occur. The throw cannot result in a 3 and a 4 at the same time. These six events are mutually exclusive.

Exclusive events can never be independent since the occurrence of one makes the other impossible and therefore has a drastic influence on its probability.

As an example consider the single throw of a fair die. The probability of a 6 is 1/6. But if a 3 occurs then the probability of a 6 is 0. The occurrence of the 3 has changed the probability of a 6 so that the events are not independent.

(b)   p (at least one 6) = 1 - p (no sixes)

p (no six on a single die) $= \frac{5}{6}$

p (n sixes on both dice) $= \frac{5}{6} \times \frac{5}{6}$ due to independence.

Hence the probability of at least one 6 $= 1 - \frac{25}{36} = \frac{11}{36}$.

(c)   p (both dice show a 6) $= \frac{1}{6} \times \frac{1}{6} = \frac{1}{36}$

If the two dice are thrown again the probability of two sixes is still 1/36 and is unaffected by the result of the first throw.

(d)   The machine will break down if at least one of the switches fails.

p (machine breaks down) = 1 - p (machine doesn't break down)

p(machine doesn't break down) = p (all 3 switches are operational)

= .9 x .9 x .9 = 0.729.

[**NB:** These probabilities multiply because switch failures are assumed to be independent]

1.3   Explanatory Notes

1.   The three definitions are self-explanatory and are concerned with fundamental concepts in probability.

2.   Relative frequency provides a useful intuitive interpretation of probability. If we say, for example, that an event has a probability of .84, then we mean that the event would occur 84% of the time if a very large number of trials were carried out. An important consequence of this is the 'expected' number of times that an event will occur. If an event has a probability of .84, and one thousand trials take place, then the 'expected' number of occurrences of the event is 1000 x .84 = 840. Note that the 'expected number' is a technical term; it does not mean that the event will occur exactly 840 times. Intuitively we may expect that the number of times it occurs will be approximately 840. It should be noted that expected numbers do not even have to be whole numbers.

3.   Although independence has an intuitively obvious meaning, it is helpful to understand a more formal definition of this concept. It can be formally defined using the probability notation. If A and B are two events with probabilities $p(A)$ and $p(B)$ respectively and if $p(A|B)$ is used to denote the probability of A given than B occurs, then A and B are independent if and only if

$$p(A|B) = p(A).$$

In other words the probability of A is unaffected by the occurrence of B. There are other formal definitions of independence and these are all equivalent to each other. For example,

$$p(B|A) = p(B)$$

and also

$$p(A \text{ and } B) = p(A) \times p(B).$$

This last definition is one of the most useful in solving problems and can be regarded as one of the fundamental probability laws. If two events are independent then the probability that both occur is simply the product of their individual probabilities. For example, the probability that two independent tosses of a fair coin result in heads is 1/2 x 1/2 = 1/4.

4.   Exclusive events are events that cannot happen together so that the probability of both happening is zero and is therefore not equal to the product of their individual probabilities; hence they cannot be independent. Students often confuse independent and exclusive events; it is important to realize that these are <u>totally</u> different properties and the existence of one denies the other.

5.   Parts (b), (c) and (d) make use of two fundamental probability laws:-

The first law states that the probability of an event is always one minus the probability that the event does not happen, (eg the probability that it rains is one minus the probability that it is dry, etc). Although intuitively obvious, students often overlook the use of this law in probability manipulations.

The second law used is that resulting from the concept of independence, namely that the probability of the occurrence of both of two independent events is simply the product of their individual probabilities. [**NB:** This does <u>not</u> hold if the events are dependent.]

1.4   Explanatory Notes (continued)

6.   Let us now see how these rules are applied in solving the problems. In (b) we are concerned with the event 'at least one 6'; the complement of this is 'no sixes'. Hence, using the first law, we can express the probability of at least one 6 in terms of the probability of no sixes. This second probability is easier to determine since it is the probability of a 'non six' on the first throw <u>and</u> a 'non six' on the second throw. Since the throws are independent then the probability of no sixes is the product of the probabilities of a 'non six' on each throw, namely

$$\frac{5}{6} \times \frac{5}{6} \; .$$

The required probability of at least one 6 is then

$$1 - \frac{5}{6} \times \frac{5}{6} = \frac{11}{36} \; .$$

7.   Part (c) tests the student's understanding of independence. Basically, if the throws are independent, then whatever happens in the past has no influence whatsoever on the future. At this point it is worth 'putting to rest' a common misconception on the 'law of averages'. Red and black numbers are equally likely on a roulette wheel; and yet if 6 reds occur in succession punters often believe that black is more likely on the next turn of the wheel saying that this must be so according to the 'law of averages'! This is absolute nonsense. The spins of the wheel are independent and the chance of a black on the next spin is not affected by the results in the past.

8.   The technique used to solve part (d) is identical to that used in solving (b). The two laws of probability are used as before.

Main Points

   •   Understand the concepts of independence and exclusive events.

   •   If two events are independent they cannot be exclusive and vice versa.

   •   The probability of an event happening is one minus the probability of its not happening.

   •   The probability of both of two independent events happening is the product of their individual probabilities.

2.1   Calculate the probability that, when two dice are rolled, the sum of the spots showing will be:

      (i)   an odd number

     (ii)   less than 9

   (iii)   exactly 12

   (iv)   exactly 4

If the pair of dice are now rolled a second time what is the probability that:

     (v)   the second roll will show the same two values as the first roll

    (vi)   the second roll will show the same total as the first roll

   (vii)   the first roll will show a total of 9 and the second roll will be less.

2.2   For ease of explanation let us assume that the dice are coloured red and blue as this will clearly not affect the probabilities.

There are 36 basic equally likely possible results since each of the 6 faces on the red die can be paired with each of the 6 faces on the blue die. These may be conveniently represented in a Venn diagram thus:-

The totals of the dice are shown next to each point:

	6	x(7)	x(8)	x(9)	x(10)	x(11)	x(12)
	5	x(6)	x(7)	x(8)	x(9)	x(10)	x(11)
Blue Die	4	x(5)	x(6)	x(7)	x(8)	x(9)	x(10)
	3	x(4)	x(5)	x(6)	x(7)	x(8)	x(9)
	2	x(3)	x(4)	x(5)	x(6)	x(7)	x(8)
	1	x(2)	x(3)	x(4)	x(5)	x(6)	x(7)
		1	2	3	4	5	6

Red Die

We now solve (i) - (iv) by counting the number of points in this diagram that correspond to the event of interest and dividing this by the total number of points, ie 36, to give the required probability.

(i)   There are 18 points corresponding to sums that are odd. Hence the probability of an odd total is

$$\frac{18}{36} = \frac{1}{2} = 0.50 .$$

(ii) There are 26 points corresponding to sums that are less than 9. Hence the probability of a total less than 9 is

$$\frac{26}{36} = \frac{13}{18} = 0.72 .$$

(iii) There is only 1 point corresponding to a sum of 12. Hence the probability of a total of 12 is

$$\frac{1}{36} = 0.03 .$$

(iv) There are 3 points corresponding to a sum of 4. Hence the probability of a total of 4 is

$$\frac{3}{36} = \frac{1}{12} = .08 .$$

2.3   For (v) - (vii) we can conceive of two Venn diagrams, one for the 36 outcomes on the first throw and one for the 36 outcomes on the second throw. Hence there are 36 x 36 basic equally likely outcomes altogether.

(v) Of these we wish to identify those that show the same faces on both occasions. There are 36 points in which the dice fall exactly the same way on both occasions and there are 30 points where the faces are the same but the colours are reversed (eg Red = 3 and Blue = 5 on the first throw, and Blue = 5 and Red = 3 on the second throw).

Hence the probability of the same faces $= \dfrac{36 + 30}{36^2} = \dfrac{11}{216} = .05$

(vi) If a particular total can occur n ways on the first throw then each of these can be associated with each of the n ways that it can occur on the second throw, making a total of n^2 possible ways. For example, from the diagram a total of 5 can occur in 4 ways for each throw and since these are associative there are 16 ways of getting a total of 5 on both throws.

Altogether, therefore, the number of ways that the totals are the same is:

$1 + 4 + 9 + 16 + 25 + 36 + 25 + 16 + 9 + 4 + 1 = 146$

Hence the probability of equal totals $= \dfrac{146}{36^2} = \dfrac{73}{648} = .11$.

(vii) From the diagram we see that the first throw can result in a total of 9 in 4 ways. The second throw can result in a total less than 9 in 26 ways. Hence the probability of a 9 on the first throw and less than 9 on the second throw is equal to

$\dfrac{4 \times 26}{36^2} = \dfrac{13}{162} = .08$.

2.4   Explanatory Notes

1. There are often a large number of correct ways of solving a probability problem. Some approaches may take longer than others but the student should not be concerned if his approach is totally different from that given here; he should be concerned if his answers are different ! This problem, for example, can be tackled using basic probability (this is the approach shown here) or by the use of various probability laws.

2. The basic probability approach can be applied when an experiment can result in several equally likely outcomes. First of all we count the number of these, N say, and then we count the number of these corresponding to the event of interest, n say. Then the probability of the event is simply n/N. This may sound very easy but care must be taken in the identification and counting of the basic outcomes.

    As a simple example let us consider tossing a fair coin twice. The possible equally likely outcomes are HH, HT, TH, TT where H = heads and T = tails. Hence N = 4. If we require the probability that the results are different, then n = 2 and so the required probability is 2/4 = .5.

3. This technique has been applied to solve all 6 sections of the dice problem. A useful device introduced here is the Venn diagram; this is simply a graphical representation of all the possible outcomes of the experiment in which each outcome is represented by a point (this set of points is sometimes called the Sample Space). For the 2 dice problem this is particularly useful and makes the points easy to count. Parts (i) - (iv) are easily solved using this diagram. Sections (v) - (vii) involve 2 throws of the dice and so the problem is a little more involved. For each throw there are the 36 equally likely outcomes shown in the diagram, but each result on the first throw can be associated with each result on the second throw so that there are, in all, 36 x 36 basic equally likely outcomes. It is not practical to represent all of these diagramatically but, with a little thought, it is not difficult to count the number of these that are relevant to the event in question and hence evaluate the probability of the event. The solutions given explain this enumeration.

4. It is instructive to look at other methods of solving some of these problems. For part (vii), for instance, we require the probability of a 9 on the first throw and a less than 9 on the second throw. Since the throws are independent this is equal to the product of the separate probabilities (see discussion of independence in Example 1 Explanatory Note 3). From the Venn diagram we see that these are 4/36 and 26/36 and so their product gives the required probability.

5. If we reconsider the solution to part (vi) then we may introduce another probability law, sometimes known as the additive law. This states that if some events are exclusive then the probability of one occurring is the sum of their individual probabilities. In part (vi) the totals are equal if they are both 2's or both 3's or ... or both 12's. These are exclusive events and so the required probability is equal to

    prob(both 2's) + prob(both 3's) + ... + prob(both 12's) .

    Now, since the throws are independent, the probability of both totals being equal to 2 is the product of the individual probabilities, namely 1/36 x 1/36. Evaluating the other probabilities in a similar manner and substituting into the above expression gives the required result.

2.5   Main Points

- When solving probability problems that involve experiments which result in equally likely outcomes (eg coins, dice, cards, random sampling) often the best approach is the basic approach.

- This involves counting the possible number of possible outcomes, counting the number of these corresponding to the event of interest, and then simply dividing one by the other.

- The Venn diagram is often a useful aid in the counting process.

- When events are exclusive (no two can happen together) the probability of one of them happening is equal to the sum of their individual probabilities.

3.1   (a)   Calculate the probabilities of drawing the following cards when 4 cards are drawn, without replacement, from a pack of 52 playing cards:

(i)   four kings,

(ii)   four picture cards (ie Jack, Queen, King),

(iii)   two hearts and two diamonds in any order.

Determine these probabilities if the cards are drawn with replacement.

(b)   Two cards are drawn at random from a pack of 52 playing cards. Calculate the probability that:

(i)   one is any heart and the other any king,

(ii)   one is the king of hearts while the other is a king or a heart.

(c)   A boy has 9 marbles - 6 red and 3 blue - in his pocket. If he extracts 3 of them what is the probability that all 3 will be red?

(d)   A box contains 12 marbles, three of each of the following colours: red, blue, green and yellow. What is the probability, if 3 marbles are drawn in succession, with replacement, that the first will be yellow, the second blue and the third green?

Assuming that the above draw produced this result, what is the probability that a further draw would produce a red marble?

(e)   Repeat the question asked in (d) but assuming that marbles are drawn without replacement.

(f)   A bag contains 5 red balls and 3 green balls. If 3 balls are drawn in succession what is the probability that the second ball drawn is red if:

(i)   sampling is with replacement,

(ii)   sampling is without replacement?

3.2   (a)   (i)   There are 4 kings and 52 cards. Hence the probability that the first card is a king is 4/52 = 1/13.

Once this has been drawn there are now 3 kings in the remaining 51 cards.

Hence the probability that the second is also a king is 3/51 = 1/17.

Therefore the probability that the first 2 cards are kings is 4/52 x 3/51.

Continuing this argument to the 4 cards gives the required probability as

$$\frac{4}{52} \times \frac{3}{51} \times \frac{2}{50} \times \frac{1}{49} = .0000037$$

(ii)   There are 12 picture cards in the pack of 52.

Hence, arguing as in (i), the required probability is

$$\frac{12}{52} \times \frac{11}{51} \times \frac{10}{50} \times \frac{9}{49} = .0018$$

(iii)   The possible orders are, (HHDD), (HDHD), (HDDH), (DHHD), (DHDH), (DDHH).

There are 13 H's, 13 D's in the pack of 52 cards.

Now $p(HHDD) = \frac{13}{52} \times \frac{12}{51} \times \frac{13}{50} \times \frac{12}{49}$

It is easily seen that each of the six orders has this as its probability.
Since the 6 orders are exclusive events then the probability of one of them is the sum of their individual probabilities (see Example 2 Explanatory Note 5).
Hence the required probability

$$= 6 \times (\frac{13}{52} \times \frac{12}{51} \times \frac{13}{50} \times \frac{12}{49}) = .0225$$

For with replacement sampling the sampled card is replaced in the pack at each stage so that we are always sampling from the same population (ie the complete pack). Hence the probability of selecting a king is always 4/52 whatever was sampled earlier.

Sections (i) - (iii) can now be solved in a similar manner but with these changes in the sampling probabilities at each stage

(i) Required probability =

$$\frac{4}{52} \times \frac{4}{52} \times \frac{4}{52} \times \frac{4}{52} = .000035$$

(ii)   Required probability =

$$\frac{12}{52} \times \frac{12}{52} \times \frac{12}{52} \times \frac{12}{52} = .0028$$

(iii)   Required probability =

$$6 \times (\frac{13}{52} \times \frac{13}{52} \times \frac{13}{52} \times \frac{13}{52}) = .0234$$

3.3   (b)   (i)   The event 'any heart and any king' can be decomposed into the 3 exclusive events:

'king of hearts and any other heart'
'king of hearts and any other king'
'king and a heart, neither of which is the king of hearts'

The probabilities of these are, respectively,

$$2 \times (\frac{1}{52} \times \frac{12}{51})$$

$$2 \times (\frac{1}{52} \times \frac{3}{51})$$

$$2 \times (\frac{3}{52} \times \frac{12}{51})$$

Hence, seeing the events are exclusive, the required probability is the sum of these giving

$$\frac{1}{26} = .0385$$

(ii)   This event is equivalent to the first two of the 3 exclusive events identified in part (i). Hence the required probability is the sum of their probabilities and is equal to

$$\frac{2 \times 15}{51 \times 51} = .0113$$

(c)   The probability that the first is red is $\frac{6}{9}$

The probability that the second is also red is $\frac{5}{8}$

etc

Hence the required probability is $\frac{6}{9} \times \frac{5}{8} \times \frac{4}{7} = .2381$

(d)   The probability that the first drawn is yellow is $\frac{3}{12} = \frac{1}{4}$

Since the marble is replaced the second draw is also from all 12 marbles under exactly the same conditions as the first draw. Hence the probability that the second marble drawn is blue is

$$\frac{3}{12} = \frac{1}{4}$$

Hence the probability of yellow, blue and green $= \frac{1}{4} \times \frac{1}{4} \times \frac{1}{4} = \frac{1}{64} = .0156$

When marbles are replaced the conditions for each draw are exactly the same each time and are clearly independent of the results of previous draws.

Hence the probability of a red marble on the 4th draw is also $\frac{3}{12} = \frac{1}{4}$.

3.4   (e)   As before the probability that the first is yellow is $\frac{3}{12} = \frac{1}{4}$.

But now there are only 11 marbles remaining and 3 blue marbles. Probability of a blue on the second draw, given that yellow was sampled on the first draw is 3/11 etc.

Hence the probability of yellow, blue, green is now $\frac{3}{12} \times \frac{3}{11} \times \frac{3}{10} = .0205$.

On the 4th drawing there are 9 marbles remaining and 3 red marbles. Hence the probability of a red, having already drawn yellow, blue, green, is

$$\frac{3}{9} = \frac{1}{3}$$

(f)   There are 8 balls in the bag and 5 of these are red.

Each of the 8 balls is equally likely to be the second ball drawn whether we are sampling with or without replacment.

Hence in both cases the required probability is

$$\frac{5}{8}$$

3.5   <u>Explanatory Notes</u>

1.   Although these exercises are all based on simple situations they are important as they demonstrate the main principles of simple random sampling with and without replacement. Of the two, sampling <u>without</u> replacement is by far the more common in practice as it is unusual, when sampling from a large group of items (known as the population), to replace sampled items and then to possibly sample the same item twice. Sampling <u>with</u> replacement is clearly less efficient and therefore less desirable.

2.   There is one, almost self-evident, point worth noting in the case of sampling without replacement; sampling 5 items, one at a time and without replacement, is <u>exactly</u> the same probability wise as sampling a group of 5 items. The main point about sampling without replacement is that at each stage of the sampling the population from which we sample changes as members of the population are removed from it. Hence the probability of sampling an item at the second stage depends on the result of the first stage. This is not the case for sampling with replacement: in this case items are replaced as they are sampled and so we sample from the same population at each stage. Hence the sampling of a particular item at any stage does not depend on the results of sampling at earlier stages so that the sampling stages are, in this case, independent.

3.   As an example of these points let us consider the problem in section (f). Here we have 8 balls, 5 of which are red and the solution states that the probability of a red ball on the second drawing is 5/8 whether we use with or without replacement. This is clear from the simple and direct argument that each of the 8 balls is equally likely to be the second ball selected. At first sight this would seem to be a contradiction of the discussion above, but it is not. If we know <u>nothing</u> about the first ball selected then the probability that the second is red is indeed 5/8. But if we <u>do know</u> the colour of the first ball selected then this changes the chances of second stage selection in the case of 'without replacement' sampling, but has no effect on the chances for 'with replacement' sampling. For example, for without replacement, let us suppose that a red ball was sampled on the first occasion, then the probability of a red ball at the second stage is now 4/7. If sampling were with replacement then the second stage probability remains at 5/8. These are the basic principles concerning with and without replacement sampling and, with a good grasp of these, the student should have no difficulty in understanding the solutions to this Example and subsequent Examples based on this type of sampling. Can you explain intuitively why, in section (a), the with replacement probabilities are higher than the without replacement probabilities?

4.   Section (b), however, may need a little more explanation. The events here are complicated and the first thing to do in such cases is to break them down into simple events that we can handle. Any breakdown will do, and yield the correct answer, <u>providing</u> the simpler events are exclusive and cover all possible ways that the event in question can happen. As a simple example of this suppose that we require the probability of an even number on a fair die. This event can be decomposed into the three events 2, 4 or 6. These are exclusive and if an even number occurs it must occur in one of these ways. Hence the probability of an even number is the sum of their probabilities. For section (b) (i) the event in question is divided into 3 simpler events and these can be seen to be exclusive and cover the event of interest. The probabilities of each of these simpler events are easily determined by the methods discussed earlier. Can you think of any other subdivision of the main event that would also work?

3.6   <u>Main Points</u>

- •   When items are sampled one at a time <u>without</u> replacement the population changes at each stage and the sampling probabilities depend on the results of previous sampling.

- •   For <u>with</u> replacement sampling the population is the same at each stage so that the probabilities at each stage are the same and independent of previous samples.

4.1   A particular audit sampling scheme involves the random sampling, without replacement, of 30 items from a total of 673 items.

A sampled item is then audited thoroughly and classified as correct or incorrect. If, in fact, 20 of the 673 items are in error determine the probability that this sampling scheme will find at least one bad item.

How many items would need to be sampled if this probability were to be raised to 0.9?

[NB: The calculations necessary to answer this question are tedious if done by hand; if no computational facilities are available then leave the answer in a form that will give the required numerical answer after computation. Better still, obtain the answers by writing a simple program either on a hand calculator or microcomputer]

4.2   Probability of sampling at least one bad item

= 1 - p (sampling all correct items)

and since sampling is without replacement then

$$p(\text{all sampled items are correct}) = \frac{653}{673} \cdot \frac{652}{672} \cdot \frac{651}{671} \cdots \frac{625}{645} \cdot \frac{624}{644}$$

= 0.3964.

Hence the probability of detecting at least one bad item = 0.6036. (ie just over 60%).

If we want this chance raised to over 90% then we must sample until the probability of observing all good items becomes less than 0.1

ie we keep adding terms to the computation $\left(\frac{653}{673}\right)\left(\frac{652}{672}\right)\left(\frac{651}{671}\right)$ ...

until this product is less than 0.1. The number of terms in the product is then the required sample size.

Computation shows that this product is less than 0.1 for the first time for 73 products.

Hence we need a sample size of 73 for the probability of detection to be as high as 0.90.

4.3   Explanatory Notes

1.   This is a prime example of a situation when it is easier to find the probability of the complement of the event of interest and take this from 1 to get the required probability.

2.   Hence we concentrate our attention on finding the probability of no bad items in the sample.

Altogether there are 653 correct items in the population of 673 items.
The probability that the first sampled item is good is 653/673.

The probability that the second sampled item is also good is 652/672 since we are now sampling from a population of 672 items containing 652 good items. This is another example of random sampling without replacement which we discussed at some length in Example 3.

Continuing this argument for 30 stages gives the probability of sampling 30 good items, and by extracting this from 1 we get the required probability.

3.   The second part of this question is based on much the same type of calculation except this time we need to determine the sample size to achieve a required probability of detection. In this case we continue to compute the probability of sampling all good items until a sample size is reached where this probability becomes less than 0.1; then the probability of detecting at least one bad item is above 0.9.

4.   This example is typical of a basic probability application to audit sampling; an applied situation may be more complicated but the same type of probability calculation, with a little modification, will apply. For example, it may be possible to divide the population of items into groups in such a way that we would wish to sample from each group and some groups may be more liable to errors than others. Then the calculation of this example would be applied separately to each group; this is known as stratified sampling.

Main Points

•   In basic probability calculations always consider finding the probability of the complement of the event of interest – it may be easier.

•   The standard sampling without replacement techniques are appropriate to audit sampling.

5.1   (a)   Components for assembly are delivered in batches of 100 and experience shows that 5% of each batch are defective. On arrival, 5 pieces are taken from each batch and tested. If 2 or more of the 5 are found to be faulty the entire batch is rejected. What are the chances that a batch will not be rejected?

(b)   It is known that 5% of the output of a given machine is faulty. What is the probability that, in a sample of 10 units taken at random from the machine, the number of faulty units in the sample would be:

    (i)   less than two,

    (ii)   exactly two,

    (iii) more than two?

(c)   Four dice are rolled and the number of 6's recorded. If the experiment is repeated 50 times on how many occasions would you expect:

    (i)   two sixes to appear, and

    (ii)   four sixes?

5.2   (a)   Each batch contains 5 faulty items and 95 good items. 5 items are sampled at random, without replacement. The probability of 2 or more defectives in the sample is equal to

$$1 - p\{0 \text{ defectives or } 1 \text{ defective in the sample}\}$$

$$p(0 \text{ defs.}) = \frac{95}{100} \times \frac{94}{99} \times \frac{93}{98} \times \frac{92}{97} \times \frac{91}{96} = 0.7696$$

$$p(\text{exactly } 1 \text{ def.}) = 5 \times \{\frac{5}{100} \times \frac{95}{99} \times \frac{94}{98} \times \frac{93}{97} \times \frac{92}{96}\} = 0.2114$$

The probability that a batch is accepted is 0.9810.

Hence the probability of 2 or more defectives = 1 - 0.7696 - 0.2114 = 0.0190.

   (b)   The probability of an item being faulty is p = 0.05.

Sample size = 10 = n.

The number of faulty units has a binomial distribution with n = 10, p = 0.05.

   (i) $p(\text{less than } 2) = p(0) + p(1) = (0.95)^{10} + 10(0.95)^9(0.05) = 0.9139.$

   (ii) $p(\text{exactly } 2) = {}^{10}C_2(0.05)^2(0.95)^8 = 45 \times (0.05)^2 \times (0.95)^8 = 0.0746.$

   (iii) $p(\text{more than } 2) = 1 - \{p(0) + p(1) + p(2)\}$

$$= 1 - 0.9139 - 0.0746 = 0.0115.$$

   (c)   The number of sixes that are recorded has a binomial distribution with

$$n = 4 \text{ and } p = 1/6 .$$

   (i) $p(2 \text{ sixes}) = {}^4C_2(\frac{1}{6})^2(\frac{5}{6})^2 = 0.1157.$

Then the expected number of occasions with 2 sixes is

$$50 \times 0.1157 = 5.79.$$

   (ii) $p(4 \text{ sixes}) = (\frac{1}{6})^4 = 0.00077.$

Expected number of occasions = 0.04.

5.3   Explanatory Notes

1.   This question is predominantly about an important probability distribution known as the <u>Binomial Distribution</u>. It applies when we have a number of trials each of which may be classified in one of two ways - let us call these 'success' and 'failure'. Then the binomial distribution is the probability distribution of the number of successes in a fixed number of trials. Examples of its application are

(a)   the distribution of the number of heads when a coin is tossed 10 times,

(b)   the distribution of the number of bad items when a sample of 50 items is taken from a production line.

For the distribution to be valid these trials must possess three important properties:

 (i)   the result of each trial is classified in one of two ways

 (ii)   the trials are independent of each other

(iii)   the probability of a success in each trial is constant and remains constant whatever the results of previous trials.

Examples (a) and (b) clearly satisfy these conditions. An important point here is that sampling without replacement does not satisfy the conditions for the binomial for, as we have seen in earlier examples, the probabilities change as we sample. If we are taking a relatively small sample from a very large population these changes in the probabilities are negligible and the binomial provides a good approximation to the true probabilities.

2.   There are two ways in which the binomial distribution may be used to give probabilities: by reading binomial distribution tables such as Table 1 or by the use of a mathematical formula. We now explain both of these methods.

3.   The Formula

Suppose we sample 8 items from a very large population that actually contains a proportion of 'success' equal to 0.3 and we wish to determine the probability that the sample contains exactly 2 successes (and therefore 6 failures). Then the binomial distribution formula gives this probability as

$$^8C_2 \qquad (0.3)^2 \qquad (.7)^6 = 0.2965$$

(A)   (B)   (C)

We now explain the components of this formula:

(A) $^8C_2$ is known as the combinatorial coefficient. It is equal to

$$\frac{(8 \times 7 \times 6 \times 5 \times 4 \times 3 \times 2 \times 1)}{(2 \times 1) \times (6 \times 5 \times 4 \times 3 \times 2 \times 1)} = 28$$

Multiplying a number by 1 less each time until we get down to 1 is known as the factorial of the number. In other words $^8C_2$ is equal to factorial 8 divided by the product of factorial 2 and factorial 6(= 8-2). If you want to check that you can evaluate such coefficients try verifying the following:

$^5C_3 = 10; \quad ^3C_1 = 3; \quad ^8C_4 = 70; \quad ^4C_3 = 4.$

5.4   Explanatory Notes (continued)

        (B) $(0.3)^2$ is the success probability for each trial raised to the power of the required number of successes.

        (C) $(0.7)^6$ is the failure probability for each trial (1 - success probability) raised to the power of the required number of failures (= number of trials less the number of successes).

The general formula for the binomial distribution is $^nC_r p^r q^{n-r}$ and this gives the probability of observing exactly r successes in n trials when the success probability for each trial is p.

4.   The Tables

[These are not provided in accountancy examinations but one is included here for general interest - Table 1 at the end of this book].

As in most statistical tables it is the cumulative probabilities rather than the individual probabilities that are tabulated. Corresponding to values of n, p and r Table 1 gives the probability of observing r or less successes. It is therefore the sum of the probabilities of 0, 1, 2, ... r successes. The individual probabilities may be obtained by differencing. To demonstrate the use of the tables let us reconsider the numerical example discussed earlier. If we look up the table with n = 8, p = 0.3, r = 2 we see that 0.5518 is the probability of observing 2 or less successes. If we look up the value corresponding to n = 8, p = 0.3, r = 1 we see that the probability of 1 or less successes is 0.2553. The difference of these, namely 0.2965, is the probability of exactly 2 successes - as before.

In this exercise we have used the binomial probability formula to solve the problems but the student may find it instructive to verify these solutions using the binomial table.

5.   Section (a) is concerned with sampling without replacement and is, strictly speaking, not binomial. We have used the usual 'without replacement' calculations to detemine the exact probability of 2 or more defectives. The sample size is small compared with the population size, however, and it is likely that the binomial will give a good approximation in this case.

Using the binomial with n = 5 and p = 0.05 we approximate the probability of 2 or more defectives by

$$1 - \{^5C_0(0.05)^0(0.95)^5 + {}^5C_1(0.05)^1(0.95)^4\} = .0226$$

and we see that, in this case, the approximation is only accurate to 2 decimal places.

6.   In section (b) we use the binomial formula wth n = 10 and p = 0.05. Note that we add the relevant individual probabilities to obtain the cumulative probabilities; ie for 'more than 2' we can add probabilities for 3, 4, 5, ... 9, 10 . But better still is to add the probabilities for 0, 1, 2 and subtract their sum from 1.

In section (c) the binomial also applies with n = 4 and p = 1/6.

5.5   Explanatory Notes (continued)

  6.   (continued)

Note that although a die can fall in 6 ways we are only interested in classifying each die as a '6' or 'not a 6' (ie 2 ways) and so the binomial applies. As we have pointed out in an earlier Example (Example 1 Explanatory Note 2) the expected number is the number of rolls multiplied by the probability. This does not have to be an integer and in general it won't be.

Main Points

  •   When a fixed number of trials are independent, can be classified in one of two ways and the classification probabilities are constant for all trials, then the binomial distribution applies.

  •   The binomial distribution gives the probabilities of successes in the fixed number of trials.

  •   The binomial probabilities may be evaluated using a formula or by the use of binomial tables.

6.1   An annual football match is played between the regulars of two local pubs, the Rose and Crown versus the Pig and Whistle. In the past the Rose and Crown have had rather the best of things and have scored an average of 1.8 goals per match; the Pig and Whistle have averaged 1.2 goals per match.

Next week the teams are to play again. Assuming that the numbers of goals scored by the teams have independent Poisson distributions, find the probabilities that the result is

   (i)   game won by the Rose and Crown

   (ii)   a draw.

6.2   (i)   A win for the Rose and Crown can occur in many ways. These possibilities may be identified as:

Possibility	Rose and Crown's score	Pig and Whistle's exact score
1	more than 0	0
2	more than 1	1
3	more than 2	2
4	more than 3	3

Applying the basic laws of probability the overall probability of a win may be expressed as:

p(win) = p(R and C scores more than 0) x p(P and W scores exactly 0)
              +
          p(R and C scores more than 1) x p(P and W scores exactly 1)
              +
          p(R and C scores more than 2) x p(P and W scores exactly 2)
              +
                              etc

As goals scored by teams follow the Poisson distribution then Poisson tables may be used to evaluate these probabilities.

For the Rose and Crown scores we refer to the Poisson tables with m = 1.8. Then, from the tables,

          p(R and C scores more than 0 goals)   = 0.8347
          p(R and C scores more than 1 goal)   = 0.5372
          p(R and C scores more than 2 goals)   = 0.2694
                        etc

For the Pig and Whistle we refer to the Poisson tables with m = 1.2

          p(P and W scores exactly 0 goals)   = 1 - 0.6988 = 0.3012
          p(P and W scores exactly 1 goal)   = 0.6988 - 0.3374 = 0.3614
          p(P and W scores exactly 2 goals)   = 0.3374 - 0.1205 = 0.2169
                        etc

These probabilities are shown in the following table:

No. of goals	0	1	2	3	4	5	6
Rose and Crown	0.8347	0.5372	0.2694	0.1087	0.0364	0.0104	0.0026
Pig and Whistle	0.3012	0.3614	0.2169	0.0867	0.0261	0.0062	0.0012

Hence the probability of a win by the Rose and Crown is

          (.8347 x .3012) + (.5372 x .3614) + (.2694 x .2169) + ...

Continuing the sum until terms become insignificant we finally have

          p(win by Rose and Crown) = 0.5144.

6.3   (ii)   The probability of a draw can be treated in a similar manner. In this case the relevant probability can be expressed as,

p(draw) = p(R and C scores 0 goals) x p(P and W scores 0 goals)
+
p(R and C scores 1 goal) x p(P and W scores 1 goal)
+
etc

From the Poisson table with m = 1.8 for Rose and Crown scores, and m = 1.2 for Pig and Whistle scores, we have

No. of goals	0	1	2	3	4	5	6
Rose and Crown	0.1653	0.2975	0.2678	0.1607	0.0723	0.0260	0.0078
Pig and Whistle	0.3012	0.3614	0.2169	0.0867	0.0261	0.0062	0.0012

Then, p(draw) = (.1653 x .3012) + (.2975 x .3614) + (.2678 x .2169) + ... etc
       p(draw) = 0.2314

6.4   Explanatory Notes

1.   There are two distinct steps involved in solving a problem of this kind. Firstly, the event of interest (eg a Rose and Crown win) has to be expressed in terms of a set of simple events. Secondly, the probabilities of these simple events are determined by the Poisson distribution and tables of this distribution can be used to evaluate these. The required probability for the event of interest can then be calculated. We have assumed that Poisson tables are available. If this is not so, then the probabilities have to be calculated from the formula:

$$\text{prob}(r \text{ goals}) = \frac{m^r \exp(-m)}{r!} \ .$$

eg $\text{prob}(\text{Pig and Whistle scores exactly 2 goals}) = \frac{(1.2)^2 \exp(-1.2)}{2 \times 1} = .2169$

2.   There are often several ways of expressing an event in terms of simpler events and any one of these methods should give the correct solution. In the solution presented here, the event 'a Rose and Crown win' has been expressed as,

'P and W scores 0 goals <u>and</u> R and C scores more than 0 goals'
or
'P and W scores 1 goal <u>and</u> R and C scores more than 1 goal'
or
etc

We could have obtained equally useful representations such as,

'R and C scores 1 goal <u>and</u> P and W scores less than 1 goal'
or
'R and C scores 2 goals <u>and</u> P and W scores less than 2 goals'
or
etc

The student is recommended to re-work the problem using this approach and verify that both approaches give the same solution.

3.   Two basic probability laws are now applied to this representation in order to express the probability of a win in terms of the probabilities of the simpler events. One law states that if events cannot happen together then their probabilities add, eg

p(P and W scores 0 goals <u>or</u> P and W scores 1 goal)
is equal to
p(P and W scores 0 goals) + p(P and W scores 1 goal).

The second law states that if events are independent then the probability that they happen together is equal to the product of their individual probabilities, eg

p(P and W scores 0 goals <u>and</u> R and C scores more than 0 goals)
is equal to
p(P and W scores 0 goals) x p(R and C scores more than 0 goals).

[NB: This assumes that the Rose and Crown's scores are completely independent of the Pig and Whistle's scores. We <u>do</u> need to assume this in order to answer the question, but this assumption hardly seems very realistic in practice].

6.5   Explanatory Notes (continued)

4.   It now remains to evaluate these probabilities using Poisson tables as we are told in the question that goals scored by a team follow the Poisson distribution. The Poisson distribution can be tabulated in three ways and different statistical tables will tabulate the Poisson by different methods. These are described in the following note. For the purposes of this solution we have used a table which tabulates the upper cumulative tails of the Poisson and this is Table 2 at the end of this book. That is, it tabulates probabilities such as:

P_6 = prob(the variable is greater than or equal to 6).

P_{10} = prob(the variable is greater than or equal to 10).

etc

The first thing we must do in using Poisson tables is to decide on the value of the Poisson parameter, m; this is simply the average number of occurrences per unit. For the Rose and Crown it is 1.8 goals per match and for the Pig and Whistle it is 1.2 goals per match.

Then we read the values of the P's from the Poisson table associated with our value of the Poisson parameter eg with a parameter value of 1.8 (value of m). We see that

P_2 = 0.5372 and P_3 = 0.2964

This means that the probability of the Rose and Crown scoring 2 goals or more is 0.5372; the probability that the Rose and Crown scores 3 goals or more is 0.2694

If p_0, p_1, p_2, ... denote the probabilities of exact scores then these may be obtained from the cumulative P's by differencing. To understand this it may be seen that

$P_2 = p_2 + p_3 + p_4 + ...$

$P_3 = p_3 + p_4 + ...$

so that

$P_2 - P_3 = p_2.$

Hence the probability that the Rose and Crown scores <u>exactly</u> 2 goals is equal to

= 0.5372 - 0.2694 = 0.2678

5.   Poisson tables are tabulated as:

Upper cumulants   eg P_6 = prob(variable is greater than or equal to 6)

Exact values   eg P_6 = prob(variable is equal to 6)

Lower cumulants   eg Q_6 = prob(variable is less than or equal to 6)

6.6   Explanatory Notes (continued)

5.   (continued)

Whatever the form of the table in use the student should be able to extract the relevant probabilities. These formats are all related in a simple fashion, eg

$$P_6 = 1 - Q_5 = p_6 + p_7 + p_8 + \ldots$$

$$p_6 = P_6 - P_7 = Q_6 - Q_5$$

$$Q_6 = 1 - P_7 = p_0 + p_1 + p_2 + p_3 + p_4 + p_5 + p_6.$$

By reading the tables as described above, the probabilities of the simple events are evaluated and the overall probabilities of a win and of a draw are easily calculated. Note that for the higher scores the Poisson probabilities are very small and can be ignored.

Upper cumulative tails of the Poisson table are given in Table 2.

Main Points

  •   The Poisson distribution often applies to random variables that are counts.

  •   The probabilities for this distribution may be determined from a formula or read from tables; there are several different formats for the tabulation.

  •   When determining the probability of a complex event try to break it down into simpler exclusive events and determine the probabilities of these.

7.1   Two per cent of the video tapes produced by a company are known to be defective. If a random sample of 100 video tapes is selected for inspection, calculate the probability of getting no defectives by using:

(i)   the Binomial distribution;

(ii)   the Poisson distribution.

7.2   The number of defectives has a binomial distribution with $n = 100$ and $p = .02$.

(i)   Hence the exact probability of no defectives is equal to

$$100_C_0(.02)^0(.98)^{100} = (.98)^{100} = .1326$$

(ii)   Since n is large and p is small the Poisson provides a good approximation to the binomial with the Poisson mean, $m = np$.

Hence $m = 100 \times .02 = 2$

Using the Poisson formula we see that the probability of no defectives is

$$e^{-m} = e^{-2} = .1353$$

This is an approximation and, as we can see, it is quite close to the exact probability of .1326.

7.3   Explanatory Notes

1. This problem demonstrates a very useful approximation to the binomial distribution. It applies when the number of trials is very large (ie about 50 or more) and when p is very close to the ends of its range (ie near to 0 or 1). As a rough guide the product np (or nq if p is near to 1) should be less than 10. When these conditions hold then the probabilities for the binomial and Poisson, using

    $$m = np,$$

    are very close and the Poisson may be used to approximate the binomial. This can be very useful for when n is large the binomial can be very tedious to compute and most binomial tables do not give values for n greater than 20. The Poisson approximation is, however, very easy to use. Note that this approximation does not apply for small n or when p is not extreme. When n is large and p is not extreme there is another useful approximation that can be applied (the normal); we shall meet this later.

2. In order to apply the approximation we simply compute m = np and then treat the number of successes as though it has a Poisson distribution with parameter m.

3. Section (i) is a straightforward application of the binomial distribution; video tapes are classified as effective or defective, the probability of each being defective is 2% and the tapes are independent. These are the correct conditions for the application of the binomial (see Example 5) with n = 100 and p = .02. Applying the binomial formula gives the exact probability.

    Section (ii) is concerned with the Poisson approximation. The correct conditions hold since n is large and p is small. We then use the Poisson with the parameter

    $$m = np = 100 \times .02 = 2.$$

    The Poisson formula is $\dfrac{e^{-m}m^r}{r}$.

    Putting m = 2 and r = 0 in this we get the approximation to the correct probability. Alternatively we could use the Poisson tables with m = 2 to determine this probability.

    Note that the <u>correct</u> distribution is the binomial and this gives the <u>exact</u> probability; the Poisson approximation is useful as it will often give a good deal of saving in tedious computation.

Main Points

- When we have a binomial distribution with a large value of n and an extreme value of p, then we will get much the same sort of probabilities, and more easily, by using the Poisson approximation.

- If p is near to zero then we use the Poisson with m = np. If p is near to one then we use the Poisson for the distribution of failures with m = nq.

8.1 A survey has been conducted on the wage rates paid to clerical assistants in twenty companies in your locality in order to make proposals for a review of wages in your own firm. The results of this survey are as follows:

Hourly rate in £

2.20	1.50	1.80	1.60	2.10
1.80	1.60	1.70	1.50	1.60
1.90	1.70	1.60	2.10	1.70
1.60	1.70	1.80	1.90	2.00

(a) Calculate the mode, median and mean wage rates.

(b) State which measure you think should be used in deciding on proposals for the new rate in your firm and why.

8.2   We first set up a frequency table for the hourly rates.

Hourly rate	Tally	Frequency
1.50	ll	2
1.60	HHl	5
1.70	llll	4
1.80	lll	3
1.90	ll	2
2.00	l	1
2.10	ll	2
2.20	l	1
Total		20

(a)   The mode is the hourly rate with greatest frequency. The mode is £1.60.

The median is the hourly rate which is such that 50% of the hourly rates are greater than it. The median is £1.70.

The mean is simply the average. We can determine this by averaging the original values or by taking the frequency weighted average in the above table.

$$\text{ie mean} = \frac{2 \times 1.5 + 5 \times 1.6 + \ldots + 1 \times 2.2}{20} = £1.77$$

(b)   The frequency distribution is clearly skewed to the low values, and so we have, as is often the case with 'positively skewed' distributions that mean > median > mode. The mode is clearly a bad measure of centrality in this case and the median is also rather low. The mean appears to be the best representative as it gives more compensation for the higher rates. Based on this mean a rate of £1.80 may be about the right level for the offer. In doing so the firm can claim to be offering a higher than average rate and a rate beaten by less than one-third of similar firms. In this way the company can be assured of a good quality clerical assistant for a reasonable expenditure.

8.3   Explanatory Notes

1.   A useful first step in describing a data set is to form some 'picture' of the distribution of the data. This is provided by a diagram known as a Histogram; this is simply the frequency distribution of the data and it is sometimes referred to as such. In this particular case the data can only take 8 values and so we count the number of times these occur. When the data can take a large number of values it is more usual to group them into intervals and associate frequencies with these intervals. This is demonstrated in Example 9. We begin, therefore, by making a frequency count.

2.   Although the question does not explicitly ask for this, the determination of the mode and median follow directly from it. We list the possible values of the wages rates and then run through the data, one at a time, to form the tally; that is we place a stroke next to the value taken by the data point. These strokes are then counted to give the frequencies associated with the set of wages rates. Note that it is conventional to place the fifth stroke across the previous four as an aid to counting the tally. The frequency diagram is then simply a plot of frequencies by wage rates. We can see immediately that the data distribution is not symmetric; there are a number of small wage rates but only a few high wage rates. A non-symmetric distribution of data such as this is said to be positively skewed.

3.   The technique known as 'descriptive statistics' attempts to summarise the main features of a data set in terms of a few relevant statistics. Two extremely important features are the location of the data set and the amount of spread about that location. We shall deal with spread later. There are several statistics that can be used to represent location and the most important and reliable of these is the simple arithmetic mean. Two others that are sometimes used are the mode and the median.

4.   The Mode is the value that occurs most. For this data set we see that 1.60 has the highest frequency and so this is the mode. Note that the mode is not necessarily unique; if there had been one more value at 1.70 then the data distribution would have been bimodal

5.   The Median is (ideally) the value that divides the data into two equal sized groups so that 50% of the data are greater than it and 50% are smaller. In practice it is often difficult to find a value which does this exactly and so we choose the value that comes nearest to satisfying the definition. For this data set we select 1.70 as the median; 7 values are smaller and 9 are larger but the division would be more unbalanced if any other value were taken. One way of looking at this is to regard the frequency at 1.70, namely 4, as divided up into '3 below and 2 above' so that 1.70 then divides the data equally.

6.   The Mean is simply the sum of the data divided by the number of data. It is unique and generally the most reliable measure of location. Note that the mean does not have to be equal to any of the data values. For example the mean of the integers 1, 2, 4 is 7/3. It makes sense to talk about a mean of $5\frac{1}{2}$ people.

7.   In the solution we have referred to the distribution as being positively skewed. The accepted terminology is to regard a distribution with a long tail to the right as having a positive skewness; a distribution with a long tail to the left has a negative skewness.

8.4   Main Points

   •   The correct first step in analysing a data set is almost always to
          look at a 'picture' of its shape using a frequency diagram or
          histogram.

   •   You should know the definitions of the three principle measures of
          location -the mean, mode and median. The simple mean is usually
          the best.

9.1   The following table shows the weekly production of items over a period of 50 weeks, by the manufacturing department of your company.

        360   383   368   365   371   382   379   372   363   350
        358   354   360   359   358   354   362   365   365   372
        364   362   385   381   393   392   395   396   396   398
        402   406   404   410   437   420   456   441   450   469
        467   460   454   459   458   453   451   444   445   438

(a)   Form a histogram of six groups to represent the distribution of weekly manufacturing production.

(b)   Construct the ogive (cumulative frequency diagram) corresponding to the histogram in (a).

(c)   Use the ogive to determine the median.

(d)   Use the ogive to determine the semi-interquartile range and comment on this and the median value.

(e)   Contrast the mean and median as measures of location.

9.2   The smallest data value is 350.
The largest data value is 469.

Then the data range is 469 - 350 = 119.

If we use 6 class intervals these will each have a width of approximately 119/6. For convenience we take intervals slightly wider than this and arrange for their division points to be at non-integers so that no data values can possibly fall on a division point.

We choose class intervals of length 20 units starting at 349.5.

Then to construct the histogram we set up the following table:

Class Interval	Tally	Frequency	Frequency
349.5-369.5	⌿⌿⌿⌿ ⌿⌿⌿⌿ ⌿⌿⌿⌿ 1	16	16
369.5-389.5	⌿⌿⌿⌿ lll	8	24
389.5-409.5	⌿⌿⌿⌿ llll	9	33
409.5-429.5	ll	2	35
429.5-449.5	⌿⌿⌿⌿	5	40
449.5-469.5	⌿⌿⌿⌿ ⌿⌿⌿⌿	10	50

(a) Histogram

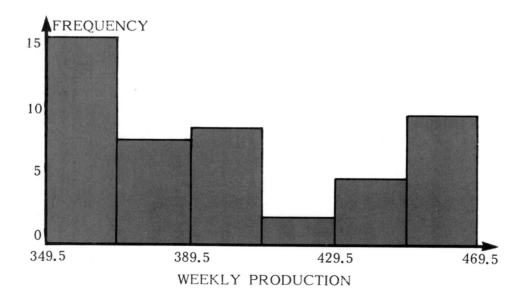

9.3   (b) Ogive

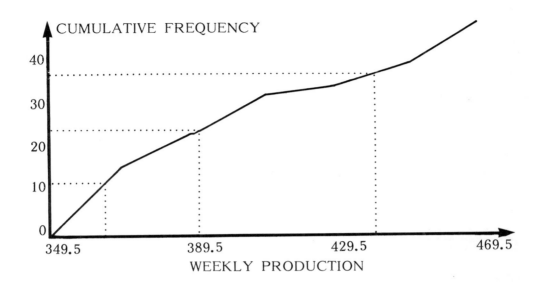

(c)   From the ogive diagram:
The median is the variable value corresponding to a cumulative frequency of 25.
This is about 390. The median is 390.

(d)   From the ogive diagram:
The lower quartile is the variable value corresponding to a cumulative frequency of
12.5. This is about 365.
The upper quartile is the variable value corresponding to a cumulative frequency of
37.5. This is about 440.
Hence the semi-interquartile range is $\frac{1}{2}(440 - 365) = 37.5$.

For a symmetric distribution the median should be about halfway between the
quartiles. In this case the mid-point between the quartiles is 402.5 which is well
above the median; this suggests that the distribution is right skewed (ie higher
frequencies at the lower values).

(e)   The histogram mean is the frequency weighted mean of the midpoints of the class
intervals, ie it is equal to:

$$\frac{1}{50} \{16 \times 359.5 + 8 \times 379.5 + \ldots + 10 \times 459.5\} = 400.3$$

This is above the median and close to the mid point between the quartiles. the
median tends to ignore the skewness but the mean takes this into account and since
the skewness is to the right the mean is greater than the median.

9.4   Explanatory Notes

1.   To construct the histogram (or frequency distribution) for a set of data such as this we start by deciding on a set of class intervals (of equal width) which cover the whole data set. The first thing to do, therefore, is to determine the smallest and largest values of the data and then make sure that the class intervals cover this data range. The size, position and number of intervals is subjective and the principal objective is to look for a sensible choice which will give a meaningful diagram. A good tip in choosing the positions of the intervals is to choose their intersections so that no data point can be equal to it; in this way each data value is allocated to one, and only one, interval.

2.   In this problem we are told to use 6 class intervals and since the data range is 119 a width of 20 for each interval, starting at 349.5, satisfies all the points mentioned above. We then 'run through' the data allocating each to a class interval and forming the tally as we do so. From this we form the frequencies and cumulative frequencies.

3.   The Histogram is then drawn as a set of blocks on the class intervals, the height of the block being proportional to the frequency. This gives a picture of the 'shape' of the data. This set of data is rather peculiar in that it is U-shaped. If we look back at the original data set we are able to see the reasons for this; it is clear that production has been increasing fairly steadily over time. Initially it was slow, then it became very fast, and then slowed down again, with the result that there are many more high and low values than middle values.

4.   The Ogive is defined to be the diagram of cumulative frequencies. We next plot the cumulative frequencies thus obtaining the Ogive. Note that the cumulative frequencies are plotted with the upper ends of the class intervals since they represent the number of data less than that value. In this case we have created the ogive by connecting the observed points by straight lines although it is often better to draw a smooth curve through them. This diagram allows us to read off (approximately) the number of data less than any value.

5.   To determine the median we use the ogive in reverse. Since there are altogether 50 data points we read off the value corresponding to a frequency of 25 ($25\frac{1}{2}$ would be more accurate but it makes little difference) and this is the median.

6.   The lower quartile is the value that divides the data in the ratio 25% below and 75% above. Hence we obtain this from the ogive as the value corresponding to a frequency of 12.5.

     The upper quartile is the value that divides the data in the ratio 75% below and 25% above, and can be read off from the ogive corresponding to a frequency of 37.5.

7.   The interquartile range is the difference between the quartiles and the semi-interquartile range is one half of this. It provides a crude measure of the spread of the data but rather exaggerates the spread in the case of a U-shaped distribution. For a more conventional distribution the semi-interquartile range would be less than one quarter of the data range but in this case it is larger.

9.5   Explanatory Notes (continued)

8.   The position of the quartiles relative to the median also indicate a slight positive skewness to the distribution.

9.   Note that we have calculated the mean from the grouped histogram data rather than the original data. This is time saving and provides a good approximation of the true data mean. We have effectively calculated the mean of 16 values at 359.5, 8 values at 379.5, etc.

The true data mean is, in fact, 399.54; this is very close to the histogram mean of 400.3.

Main Points

•   The histogram is an extremely useful method of presenting the shape of a data set so that its main features can immediately be noticed by eye.

•   The ogive is useful in estimating percentages of data less than a certain value.

10.1   A random sample of 18 invoices from an audit have the following values (in £):

26   45   27   67   59   36

80   64   23   17   63   59

81   57   60   56   55   ·55

Calculate the mean, variance and standard deviation of this sample. Estimate a range of values which should contain about 95% of the values of all invoices in the audit; what assumption have you made in making this calculation?

10.2   We first compute some relevant statistics from the data. These are:
           number of data, n = 18
           data total = 930
           total sum of squares = 54,056

Then the sample mean is equal to $\frac{\text{Total}}{n} = \frac{930}{18} = 51.67$

We next compute the corrected sum of squares for these data.

Corrected sum of squares = $54{,}056 - \frac{(930)^2}{18} = 6{,}006$

Then the sample variance = $\frac{6006}{18-1} = 353.29$

The sample standard deviation is the square root of this so that it is equal to 18.80 and is associated with 17 degrees of freedom.

Roughly speaking 95% of the population of invoice values lie within two standard deviations of the mean. Hence the estimated range is

        $51.67 \pm 2 \times 18.80$

        giving   (14.07, 89.27)

This estimate will be quite good provided the distribution of all the invoices in the population is close to an important standard distribution known as the Normal Distribution. We shall discuss this distribution in Examples 12, 13, etc.

10.3   Explanatory Notes

1.   We have already met the sample mean as a measure of location. We now meet the most important measure of the spread of a data set about its mean; this measure is a statistic known as the <u>Standard Deviation</u>. We may think of the calculation of this in several steps.

   <u>Step I</u>   Calculate the data total

   <u>Step II</u>   Calculate the sum of the squares of the data.
   This is known as 'the raw sum of squares'.

   <u>Step III</u> Square the total and divide this by the number of data.
   This value is known as 'the correction for the mean'.

   <u>Step IV</u>   Take the correction for the mean away from the raw sum of squares.
   The resulting calculation is known as 'the corrected sum of squares'. It is a very important statistic, especially in correlation and regression problems.

   <u>Step V</u>   Divide the corrected sum of squares by (n-1) where n is the number of data values. Note that we divide by (n-1) not n.
   The resulting calculation is known as 'the sample variance'.

   <u>Step VI</u>   The square root of the sample variance is the 'standard deviation' of this data set.

   There is clearly a lot of jargon involved in this calculation; these terms are important and the student should become familiar with this nomenclature.

2.   The formula is best remembered in word form as:

   $$\text{Corrected sum of squares} = \text{raw sum of squares} - \frac{(\text{total})^2}{n}$$

   This is then divided by (n-1) to give the sample variance and the square root of this is the standard deviation. There are several other equivalent ways of calculating the standard deviation; for example we have,

   $$\text{Corrected sum of squares} = \text{raw sum of squares} - n \times (\text{sample mean})^2$$

   or

   $$\text{Corrected sum of squares} = \text{sum of squares of corrected data}$$

   ie the sample mean is subtracted from each data value to give a corrected data value and these are then squared and summed.

3.   Note that the 'corrected data' must always sum to zero as they are simply deviations from their own sample mean. Effectively, therefore, there are only (n-1) independent values since, if we are told (n-1) of them, then the nth is determined because the total is zero. This is the reason why we divide by (n-1) to obtain the sample variance; we call (n-1) the <u>Degrees of Freedom</u> of the sample variance. We shall discuss this in more detail in later Examples. A good habit is to always state the degrees of freedom when presenting a sample variance or sample standard deviation as the degrees of freedom represent the accuracy of this statistic and are important in inferences based upon it.

4.   Easily the best way to compute a standard deviation is to use a pocket calculator! Many pocket calculators have a standard deviation button and all one needs to do is to type in the data and then press it.

10.4   Explanatory Notes (continued)

5.   One word of warning: sometimes books and even calculators define sample variance as the corrected sum of squares divided by n rather than (n-1). This definition has no application and has to be multiplied by a correction factor for application to bring it in line with the one given here. Fortunately this usage is rare but the student should be on guard and not be confused if a text presents it in this way. To test which definition a calculator is using try typing in the data set (1, 2, 3). Using our definition the standard deviation is 1. Using the other definition the standard deviation is 0.816. The student should calculate a few standard deviations to ensure understanding of the formula and to achieve a familiarity and confidence with this important statistic.   Try verifying the following:

Data	Standard Deviation
1, 2, 3	1
8, 9, 8, 10, 12, 11, 7	1.80
1.2, 1.3, 1.6, 2.1, 1.5, 1.3	.33

6.   We have said that the sample variance and sample standard deviation are measures of spread. This can be seen directly from the formulae from which they are calculated but is more obvious from the last part of the question. Here it is stated, and we shall give more justification to this in later examples, that approximately 95% of the population lies within 2 standard deviations of the mean.

7.   It is clear, therefore, for a widely spread population the standard deviation would have to be large since the interval needed to include 95% of such a varied population would have to be wide. For a very concentrated population, however, 95% of the population would be contained in a short interval and the standard deviation would be small.

Main Points

    •   There is only one main point to this problem and that is for the student to become familiar with the calculation and concept of variance and standard deviation.

11.1   The following table shows the age distribution of females in England and Wales in mid-1981:

age in years	population in millions
0-4	1.7
5-14	3.4
15-24	3.8
25-34	3.5
35-54	5.8
55-74	5.5
75-94	2.0
above 94	negligible

It is assumed that ages are reported as integers (whole numbers of years) and it is the statistics for these integer data that we require.

(a)   Calculate the mean and standard deviation for ages.

(b)   Explain whether a sample of 400 females, sampled at random, whose mean age was 37 years, is representative of the female population for this characteristic. What explanation can you offer for a difference between the sample and population means?

11.2   This is a population frequency distribution rather than a sample distribution.

To determine the mean and standard deviation from this form of the data we treat each class mark (ie mid point of a class interval) as though it were a data value occurring with the associated frequency.

ie $\dfrac{0 + 4}{2}$ occurs 1.7 million times

$\dfrac{5 + 14}{2}$ occurs 3.4 million times

etc.

Total number of females = 25.7 million

Age total = $(2 \times 1.7) + (9.5 \times 3.4) + (19.5 \times 3.8) + ... + (84.5 \times 2)$

= 994.9 million years

Population Mean age $= \dfrac{994.9}{25.7} = 38.7$ years

Raw sum of squares = $\{2^2 \times 1.7\} + \{(9.5)^2 \times 3.4\} + ... + \{(84.5)^2 \times 2\}$

= 53,451.8 (million years)^2

Corrected sum of squares = $53,451.8 - \dfrac{(994.9)^2}{25.7} = 14,937.169$

Population variance = $14,937.169 / 25.7 = 581.213$

(numbers are so large dividing by n or n-1 makes no difference)

Population standard deviation = 24.1 years.

The sample mean of 37 is very close to the population mean in view of the large variability of the population. Admittedly we would not expect a sample mean of a sample as large as 400 to exhibit as large a variability as the basic ages; in fact we shall see in Example 14 that the standard deviation of this sample mean is

$\dfrac{24.1}{\sqrt{400}} = 1.205$

so that it is within two standard deviations of the population mean. It would appear, therefore, that this sample is representative.

11.3   Explanatory Notes

1. This question demonstrates that the mean and standard deviation can be used to represent location and spread respectively for samples or for the whole population. In this problem we are dealing with the total population of females and the means and standard deviation are fixed quantities - often referred to as the population Parameters.

2. Often in practice we do not have the values for the whole population and have to make do with a sample from it. It is important to specify how this sample is taken. If it is a random sample it is chosen in such a way that every member of the population has the same chance of being included in the sample. Once we have the sample we use it to draw inferences about the whole population and this is one of the main purposes of statistical methodology; namely to draw conclusions about the whole from the part. Such inferences will clearly be subject to error and we use probability to assess the chances of error. From a sample we can calculate a histogram, ogive, mean and standard deviation. Each of these can be regarded as estimating its population counterpart. The histogram gives some idea of the distribution of the whole population. The ogive can be used to estimate the percentage of the population in certain categories. The sample mean and sample standard deviation are estimates of the population mean and population standard deviation. Make sure you distinguish between sample statistics and population parameters. In this problem the population mean age is 38.7 and is fixed. The sample mean for the random sample of 400 females is 37. If a second sample of 400 is taken this will have a different sample mean, 39.2 say. The point here is that sample means vary due to sampling variability but the population mean is fixed.

3. These are basic but very important concepts. They form the foundation of statistical inference techniques and a clear grasp of them at this stage will enhance the understanding of later ideas.

4. The calculation of the mean and standard deviation for the population is straightforward as the same formulae are used for samples and populations. The only slight complication is that the calculations have to be performed on grouped data rather than the raw data (which would have been very tedious!). The technique is to regard this presentation as a set of basic data when each class mark (mid point of a class interval) occurs with its associated frequency. Then the formulae apply in the usual way.

Main Points

*   The mean, variance and standard deviation calculations can be applied to the whole population. They are then known as population parameters.

*   It is important to distinguish between the sample statistics and the population parameters. The sample mean and sample standard deviation can be regarded as estimators of the population mean and population standard deviation.

12.1   The variable Z has a standard normal distribution.

Determine the following probabilities:

(a)   Z is greater than 1.00

(b)   Z is greater than -1.80 but less than 0.50

(c)   Z is less than 2.40

(d)   Z lies between 1 and 2

Find the value of a constant a so that the following probabilities are correct:

(e) $p(-a < Z < +a) = 0.95$

(f) $p(-a < Z < +a) = 0.99$

12.2   (a) $p(Z > 1) = 0.5 - 0.3413 = 0.1587$

      (b) $p(-1.8 < Z < 0.5) = 0.1914 + 0.4641 = 6556$

      (c) $p(Z < 2.4) = 0.5 + 0.4918 = 0.9918$

      (d) $p(1 < Z < 2) = 0.4772 - 0.3413 = 0.1359$

      (e) $p(-a < Z < +a) = 0.95$

           Hence, by symmetry, $p(0 < Z < +a) = 0.475$

           Hence $a = 1.96$

      (f) $p(-a < Z < +a) = 0.99$

           Hence, by symmetry, $p(0 < Z < +a) = 0.495$

           so that $a = 2.57$

12.3   Explanatory Notes

1.   This is a purely technical question on the handling of the standard normal distribution. Familiarisation with this distribution and its tabulation are absolutely necessary for understanding and using the general normal distribution and its applications. The relevant table is Table 3 at the end of this book; this table is supplied as part of several accountancy examination papers.

2.   The standard normal distribution looks like this:

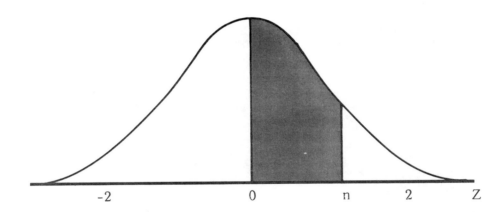

The values along the horizontal axis are those of the variable Z. The values in the vertical direction do not have an obvious direct interpretation but the important feature is that the <u>area</u> underneath this curve represents <u>probability</u>; by this we mean that the area under the curve between Z = 1 and Z = 4 is equal to the probability that Z falls between 1 and 4.

12.4   Explanatory Notes (continued)

3.   The main features of this distribution are:

(a)   It is symmetric about its population mean which is 0

(b)   Area under the curve represents probability

(c)   The total area under the curve is equal to the total probability and is therefore equal to 1.

4.   These properties, together with the table, enable probabilities to be determined for any statement about Z.   The Standard Normal Table (Table 3) tabulates the area (probability) under the curve between 0 and n, and for values of n ranging from 0 to 3.09.   There is no need to tabulate the distribution for negative values of n since these can be deduced using the symmetry property.   Tables also exist for values of n greater than 3.09 but there is very little probability beyond this point.

5.   The best way to demonstrate this technique is by example and so we now give a detailed solution to part (b).   It is not suggested that the student writes solutions in such detail as this; this is for explanatory purposes only.

For section (b) we require $p(-1.8 < Z < 0.5)$

Step 1:

Draw a 'thumbnail' sketch of the standard normal and identify, on the z axis, the interval of interest.   This is the interval (-1.8, 0.5).

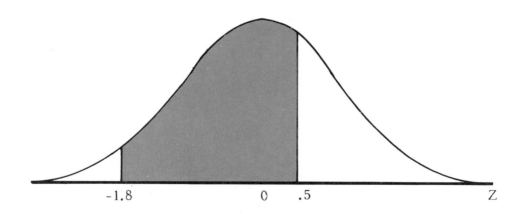

12.5   <u>Explanatory Notes</u> (continued)

5. (continued)

<u>Step 2</u>
Shade in the area under the curve on the interval of interest.

<u>Step 3</u>
Decide the best way of finding this area from the tables. This often involves breaking it up into smaller areas. In this case we divide it into areas A and B.

<u>Step 4</u>
Read appropriate areas (probabilities) from the tables.
Area A can be read directly from the table corresponding to n = 0.5.
A = 0.1915

Area B is, by symmetry, equal to the area between 0 and + 1.80.
This is the table reading for n = 1.80.
B = 0.4641.

Then the required probability is 0.1915 + 0.4641 = 0.6556.

Using this technique the student should have no trouble in understanding the solutions to the other sections of the question. As he becomes more proficient in dealing with the normal distribution he should not need to draw 'thumbnail' sketches, but this is a good aid when mastering this technique for the first time.

There are a few other points that may need clarification.

6.   We have referred to zero as the <u>Population Mean</u> of the standard normal distribution. It is well known that if we calculate a sample mean for an ever increasing sample size then the sample mean eventually 'settles down' to a constant; it is this constant that we call the population mean of the distribution. If we took an enormous sample (millions, say) of Zs from a population which was distributed as the standard normal then this mean would be zero.

7.   We have not discussed the probabilities of Z being exactly equal to some value, 1.4 say. This is, in fact, zero since there is no area on a point (zero interval). This makes sense since there is no chance of getting <u>exactly</u> 1.4. If we really mean 1.4 to the nearest decimal place then we are really talking about the interval 1.35 to 1.45 and the usual method for determining the probability for an interval applies.

8.   There is no need to understand the mathematics of the normal distribution but for those with mathematically inquisitive minds the following may be of some interest: the mathematical function that defines the standard normal is $(2\pi)^{-\frac{1}{2}} \exp(-\frac{1}{2}Z^2)$. It is defined over the whole interval $-\infty$ to $+\infty$, but for all intents and purposes the interval is -3 to +3 as there is so little probability outside the interval.

<u>Main Points</u>

*   The standard normal is a bell-shaped distribution which is symmetric about 0.

*   Area under its curve represents probability and the total area is 1.

*   Its tabulation records the area (probability) between 0 and n.

13.1   (a)   What is the Normal Distribution? What is its significance for statistical sampling?

(b)   Your company requires a special type of light bulb which is available from only two suppliers. Supplier A's bulbs have a mean lifetime of 2000 hours with a standard deviation of 180 hours. Supplier B's bulbs have a mean lifetime of 1850 hours with a standard deviation of 100 hours. The distribution of the lifetimes of each type of bulb is normal.

Your company requires that the lifetime of a bulb be not less than 1500 hours.

All other things being equal, which type of bulb should you buy, and why?

(c)   An interviewer in a particular survey is allowed 30 minutes for each questionnaire. In practice the average time is 35 minutes; the times taken are normally distributed about the mean time with a standard deviation of 5 minutes. What proportion of questionnaires would you expect to be completed within the time allowed?

(d)   30% of a certain population is estimated to have an IQ of 100 or less. A further 10% is estimated to have an IQ of 140 or more. Assuming that the characteristic of intelligence is normally distributed, what is the mean and standard deviation of this population?

13.2   (a)   The normal distribution is a bell-shaped distribution that applies to variables in many practical situations. Peoples heights and weights, petal widths of flowers, errors of measurement etc. are all found to have normal distributions. There are many normal distributions each being characterised by its population mean and population variance. The distribution is symmetric about its (population) mean and about 95% of the probability lies within two standard deviations of the mean. The normal distribution plays a very important role in sampling theory; providing the sample size is reasonably large the sample mean will have (approximately) a normal distribution. This property underlies many important inference techniques.

(b)   Let X = the lifetime of a bulb manufactured by supplier A.

Let Y = the lifetime of a bulb manufactured by supplier B.

Then X is $N\{2000,(180)^2\}$ and Y is $N\{1850, (120)^2\}$

Probability that a bulb from A is satisfactory = $p(X > 1500)$

$$= p(Z > \frac{1500 - 2000}{180})$$

$$= p(Z > -2.78) = 0.9973$$

Probability that a bulb from B is satisfactory = $p(Y > 1500)$

$$= p(Z > \frac{1500 - 1850}{120})$$

$$= p(Z > -2.92) = 0.9982$$

We should favour the bulbs from supplier B since they have a higher probability of being satisfctory (although the difference is small). In a large batch, therefore, we should expect more satisfactory bulbs from supplier B. Supplier B's bulbs are also less variable (a standard deviation of 120 as opposed to a standard deviation of 180) and so a batch from supplier A would contain a lot more 'very bad' bulbs than a batch from supplier B.

(c)   Let T be the length of an interview

Then T is $N\{35,25\}$

The probability that a questionnaire will be completed within 30 minutes is

$$p(T < 30) = p(Z < \frac{30 - 35}{5})$$

$$= p(Z < -1) = 0.1587$$

This is the proportion of interviews that will be completed in the allowed time.

(d)   Let X = the IQ of a member of the population.

Let us assume that X is $N(\mu,\sigma^2)$

Then $p(X < 100) = 0.30$

so that $p(Z < \frac{100-\mu}{\sigma}) = 0.30$

and from the standard normal table it follows that

$$\frac{100-\mu}{\sigma} = -0.525 \qquad\qquad (1)$$

13.3   Also we have $p(X > 140) = 0.10$

so that $p(Z > \dfrac{140 - \mu}{\sigma}) = 0.10$

Hence $\dfrac{140 - \mu}{\sigma}) = 1.28$ (2)

Solving the simultaneous equations (1) and (2) we have

The population mean IQ is 111.6

The population standard deviation is 22.16.

Explanatory Notes

1.   This question demonstrates the application of the general normal distribution. There are many normal distributions and they all have the shape of a bell, but they differ from each other in the point about which they are centred, and their spread about that centre point. Each normal distribution is completely defined by specifying its population mean and population variance (or standard deviaton).

2.   To understand the meaning of these population quantities let us consider the following. In previous examples we have seen how to calculate histograms, means, variances and standard deviations from samples. These are known as Sample statistics. This could be done, conceptually, for larger and larger samples until the sample sizes are huge; when this happens the statistics, which may vary a great deal in small samples, settle down. The histogram becomes a smooth curve known as the probability distribution of the population. The sample mean settles down to a constant known as the Population mean. The sample variance and sample standard deviation settle down to the population variance and population standard deviation respectively. Be sure to distinguish between the sample statistics on one hand and the population values on the other. Sample statistics vary from sample to sample but population values, often called parameters, are constant. For the general normal the population mean represents its centrality and the distribution is symmetric about this point. The population standard deviation, which is the square root of the population variance, represents the spread of the distribution, this may be directly interpreted by saying that approximately 95% of the probability content lies within two standard deviations of the mean.

3.   A normal distribution is completely specified, therefore, by a statement such as 'the variable X has a normal distribution with a mean equal to 5 and a variance equal to 4'. Instead of having to repeat this wordy sentence each time, we write this in the more convenient form

    X is N(5,4)

Note that it is the variance, and not the standard deviation which is used in this expression so that if X had a normal distribution with a mean μ and a standard deviation σ we would write

    X is $N(\mu, \sigma^2)$

13.4   Explanatory Notes (continued)

4.   The standard normal tables which were described in the previous Example can be used to determine probabilities for the general normal.   To do this we must standardise the value of interest.   Standardising consists of 'taking away the mean and then dividing by the standard deviation'.   A comparison of X with a value is exactly the same as the comparison of Z (the standard normal) with the standardised value.   Let us see how this works by considering part (b) of this example.   We are told that the lifetime X of a bulb from supplier A has a normal distribution with mean 2000 and a standard deviation of 180.   In our notation we simply write this as

$$X \text{ is } N\{2000, (180)^2\}$$

We now require the probability that X exceeds 1500.

We write this as $p(X > 1500)$

Now the comparison of X with the value 1500 is exactly the same as the comparison of Z with the standardised value of 1500; to get this value we deduct 2000 and then divide by 180.   Hence we have

$$p(X > 1500) = p(Z > \frac{1500 - 2000}{180})$$

We can now find this probability for Z from the standard normal tables - just as we did in the previous example.

This simple technique of replacing a probability statement about X by a probability statement about Z is used to solve the remaining sections of this example.   Once the notation and standardisation are mastered, there should be no difficulties in following these solutions.

Main Points

· The general normal distribution is defined by the values of the mean and standard deviation and is represented by the notation $N(\mu, \sigma^2)$.

· The standardisation operation translates a probability statement about a general normal variable to that about the standard normal, for which we have tables.

14.1   A random sample of 25 audits is to be taken from a company's total audits, and the average value of these audits is to be calculated.

Explain what you understand by the sampling distribution of this average and discuss its relationship to the population mean.

Is it reasonable to assume that this sampling distribution is normal?

If the population of all audits has a mean of £54 and a standard deviation of £10 find the probability that

(a)   the sample mean will be greater than £60

(b)   the sample mean will be within 5% of the population mean,

14.2   When we have sampled the 25 audits we total them and divide by 25 to calculate their sample mean. If we now sampled a further 25 audits and repeated this calculation we would get a second sample mean which would differ from the first. Conceptually we could imagine repeating this process of sampling and calculation several times; each time we would get a different value for our calculation. The sample mean is varying from sample to sample due to sampling variability. If we sampled a huge number of such means and drew their histogram we would have a picture of their distribution; it is this that we call the sampling distribution of the sample mean and we can use it to make probability statements about the sample mean.

If we calculated the sum of all audits and divided by the total number of audits we arrive at the <u>Population Mean</u>. This is a fixed quantity and, unlike the sample mean, it does not vary from sample to sample. In practice, however, we rarely know the population mean and we regard the sample mean as an estimator of it. It is very important to distinguish between these two types of mean; a useful view of their relationship is to note that if we were to take a larger and larger sample size then the sample mean becomes the population mean as the sample becomes the population.

Providing we use a reasonable sample size (20 or more) then it is a fact that the sample mean is approximately normally distributed. The approximation is generally very good and so it can be treated as normal for practical purposes. If the sample size is small then the sampling distribution will be approximately normal only if the distribution of the basic variable is approximately normal.

If $\bar{x}$ represents the sample mean of 25 randomly sampled audits then,

$$\bar{x} \text{ is distributed as } N\{54, \frac{100}{25}\}$$

ie $\bar{x}$ is distributed as $N(54, 4)$

(a) $p(\bar{x} > 60) = p(Z > \frac{60 - 54}{2})$

$$= p(Z > 3) = 0.5 - 0.4987 = 0.0013$$

(b)   5% of the population mean $= .05 \times 54 = 2.7$

Hence we require

$$p\{51.3 < \bar{x} < 56.7\}$$

$$= p\{\frac{51.3 - 54}{2} < Z < \frac{56.7 - 54}{2}\}$$

$$= p\{-1.35 < Z < 1.35\}$$

$$= 2 \times 0.4115 = 0.8230$$

14.3   Explanatory Notes

1.   In this example we meet the very important concept of a sampling distribution for a statistic. For the sample mean, in particular, there is a very important and well known property; it says that providing our sample size is reasonably large the sampling distribution of the sample mean is, for all practical purposes, normal. This is an important robustness property for it means that we may treat the sample mean as having a normal distribution whatever the probability distribution of the basic variable. A very useful, and somewhat surprising property!

2.   A more detailed statement of this result, that allows us to determine probabilities for the sample mean, is the following:

   If a random sample of n (fairly large) values is drawn from a population that has a (population) mean μ and a (population) variance σ^2 then the sampling distribution of the sample mean may be taken to be

   $$N(\mu, \frac{\sigma^2}{n}).$$

   Note that the variance of the sample mean is an nth of the variance of the original population so that as the sample size gets bigger the sampling distribution concentrates more and more about μ until, for huge n, $\bar{x}$ becomes μ since the variance becomes zero. This explains why larger sample sizes give more reliable estimators of μ; $\bar{x}$ has a higher chance of being close to μ.

3.   The standard deviation of the sample mean is $\sigma/\sqrt{n}$. To distinguish this from the standard deviation of the basic population, namely σ, it is often called The Standard Error.

4.   Let us now see how to apply this sampling distribution to find probabilities for $\bar{x}$. Let us consider section (a) of the example.

   We are told that $\mu = 54$ and $\sigma = 10$.

   Hence we may state that, since $\bar{x}$ is the sample mean of 25 values,

   $$\bar{x} \text{ is distributed as } N(54, \frac{100}{25})$$

   We now require $p(\bar{x} > 60)$ and, as before, the comparison of $\bar{x}$ with 60 is exactly the same as comparing the standard normal, Z, with the standardised value of 60; but we must standardise 60 using the appropriate parameters of the distribution of $\bar{x}$. We reduce 60 by the mean of the sampling distribution and then divide it by the standard error of the sampling distribution.

   Hence we have

   $$p(\bar{x} > 60) = p(Z > \frac{60 - 54}{10/5}),$$

   and this probability can now be determined from standard normal tables.

Main points

-    When taking a fairly large sample the sample mean may be treated as having a normal distribution even when the original population is anything but normal.

-    If μ and σ are the population mean and standard deviation respectively of the sampled population then the distribution of a sample mean of n values is

15.1   Describe and compare the following three methods of sampling in an audit sampling context:

(a)   Simple Random Sampling

(b)   Stratified Random Sampling

(c)   Systematic Sampling

15.2   (a)   Simple random sampling is the most basic sampling technique and is conducted in such a way that each item has an equal chance of being sampled. If, for example, there are 5000 items altogether and a simple random sample of 80 is taken, then the chance of a particular item being included in the sample is 80/5000. Also the chance of an item being selected is completely independent of whichever items have already been selected.

There are a number of ways of physically selecting a simple random sample. Among the most common is the method of stepwise selection using random numbers. Each of the total population of items is allocated an identification number (from 1 to 5000 say) and a random number is then taken from the range 1 to 5000 and the corresponding item is included in the sample. A second random number is then chosen from 1 to 5000 and the corresponding item added to the sample; this is continued until the sample consists of 80 different items. If, at any stage, the random number chosen coincides with a previous random number then it is ignored. This type of sampling is, therefore, sampling without replacement; this was discussed in Examples 3 and 4. Random numbers may be generated from computers, calculators or tables. Whichever way you select them it is important to ensure that each number in the range of interest has an equal chance of being chosen.

The principal advantage of simple random sampling is its fairness - or more technically, its unbiasedness. An important criterion for any sample is that it should fairly represent the population and not be dominated by an extreme part of the population; the method of using random numbers to select the sample members guarantees this fairness.

Once the sample has been selected the chosen items are thoroughly audited and the results of this analysis are known as the sample values; these are then used to estimate aspects of the total population of items such as total value, proportion containing errors, etc. An example of the calculations involved is given in Example 16.

(b)   We can improve on the technique of simple random sampling, and still retain its fairness (unbiasedness), when we are able to identify a natural grouping of the total population of items. This grouping should be such that items are similar within groups but different from items in other groups. These groups are known as Strata. A sample is then taken by applying simple random sampling to each of the strata separately. The sample sizes for the strata should depend on the number of items in a stratum and the variability of the items in a stratum. If the number of items in a stratum is large then a large sample size should be used. If the items in a stratum are very variable (in terms of the measurement of interest, eg value) then a large sample size should be used. For a small stratum with little variability a small sample size will be sufficient.

This type of sampling is known as Stratified Random Sampling; it retains the unbiasedness of simple random sampling but generates more accurate estimators.

In an audit sampling exercise, for example, in which the main objective is to estimate the proportion of items in error, it is better to sample more intensely from a group which is known to be prone to errors than from a group which is expected to be error free.

15.3 (c) <u>Systematic Sampling</u> applies when the items to be sampled have some natural ordering, either by some identification number, as is likely to be the case in audit sampling, or by some geographical layout. The technique consists of sampling items at equidistant points along the sequence. If, for example, we are to sample 80 items from 5000 then we could sample the first item at random from the first 80 items and then every 60th item thereafter.

The main advantage of this method is that it is simple and less time consuming than simple random sampling but it may easily result in a biased sample. If, unknown to the user, large valued items occurred at regular intervals in the sequence and these coincided with the sampling intervals, then the sample would include a far higher proportion of these than in the totality of items. This bias would lead us to believe that the total value is much higher than it really is.

Explanatory Notes

1.   This solution is substantially self-explanatory. Another important property of simple random sampling, apart from its fairness, is that standard statistical inference methods such as confidence intervals and hypothesis testing are relevant to the data that result from this type of sampling. This is also true of stratified sampling. It is wrong to apply such inference techniques to sampling techniques which do not have a probability basis; quota sampling for example. They may be applied to systematic sampling but care should be taken to avoid the biases that can result from this sampling method.

2.   Stratified sampling is a commonly used method as it retains the fairness of simple random sampling but increases the accuracy when there are some obvious strata. The choice of sample sizes for the strata is largely subjective but if the user follows the guidelines on size and variability that are discussed in the solution then a sensible sampling scheme will result.

Main Points

- Simple random sampling is fair and is the basis for most good sampling schemes.

- Stratified random sampling is a powerful and commonly used technique; it will often be appropriate to audit sampling.

- Systematic sampling is often simple and economic to perform but the results should be treated with caution since bias can often result.

16.1   A company's existing stock holding consists of 5000 items and it is necessary to make a quick estimate of the total value of this holding.

The items vary in value but may be grouped into three price categories as follows:

About 20% of the stock is comprised of low price items

About 50% of the stock is comprised of middle price items

About 30% of the stock is comprised of expensive items

Give an account of the points you would take into consideration in the design of your sample, and of the calculations you would perform to estimate the total stock value from the values of the sampled items.

16.2   It is implied that there is not time to examine all 5000 items and a sample must be taken. The first consideration, therefore, is the size of the sample and this is dependent on the time available; this would need to be known and the size of the sample would be the maximum that this time period would allow. Let us assume that this is 100.

The totality of items clearly divides into three strata and a good scheme will be to use stratified sampling; ie to use simple random sampling within each of the strata. We then have to decide how to divide the sample size of 100 among the strata. In deciding on the strata sample sizes we should take into account (a) the sizes of the strata that are given, (b) the variabilities of the strata - this often requires subjective assessment.

Let us suppose that, after discussions with employees who have handled past data of this type, we assess that the variabilities (standard deviations) are roughly in the ratios

2 : 4 : 5 for low/middle/expensive

In a practical situation it is likely that the more expensive items are more variable in price than the less expensive.

Now the strata sizes are in the ratios

2 : 5 : 3 for low/middle/expensive

Then our strata sample sizes should be chosen with ratios proportional to the <u>products</u> of these two sets of ratios

ie   2x2 : 4x5 : 5x3 for low/middle/expensive

Hence the strata sample sizes are

$$100 \left(\frac{4}{4 + 20 + 15} \right) = 10 \text{ for low price items}$$

$$100 \left(\frac{20}{4 + 20 + 15} \right) = 51 \text{ for middle price items}$$

$$100 \left(\frac{15}{4 + 20 + 15} \right) = 39 \text{ for expensive items}$$

We now sample 10 items from the low price items using simple random sampling, 51 from the middle price items and 39 from the expensive items.

We estimate the total stock value in the following manner:

Since 20% of the 5000 items are low price then there are 1000 low price items.

If we total our sample of 10 low price items then this must be multiplied by 1000/10 = 100 to estimate the total for all low price items.

Since 50% of the 5000 items are middle price, then there are 2500 middle price items.

If we total our sample of 51 middle price items then this total must be raised by 2500/51 to estimate the total for all middle price items.

Since 30% of the 5000 items are expensive then there are 1500 expensive items altogether.

If we total our sample of 39 expensive items we then multiply this by 1500/39 to estimate the total value of the expensive items.

The total stock value is then estimated by the total of these three estimates.

16.3   Explanatory Notes

1.   This is a standard stratified sampling situation. If simple random sampling had been applied to the whole stock holding, ignoring the strata, then a much less accurate estimator of the total turnover would have been obtained.

2.   The solution given is rather sophisticated in that it attempts to take the variabilities of the strata into account in deciding on strata sample sizes. This often has to be done on a subjective basis in the hope that 'some attempt is better than no attempt'.

3.   A simpler approach is to choose the strata sample sizes to be proportional to the strata sizes so that in this example we would have used 20, 50, 30 as the sample sizes. This is known as proportional stratified sampling and it is a very reasonable approach; in most cases it will only be slightly inferior to the solution presented here and will be very superior to simple random sampling from the total population of items.

Main Points

•   Use Stratified Random Sampling when the population divides into natural strata.

•   Think carefully about the choice of strata. A reasonable choice based on these principles will always result in a good sampling scheme.

17.1   A business requires an inexpensive check on the value of stock in its warehouse. In order to do this, a random sample of 50 items is taken and valued. The average value of these is £320.41 with a standard deviation of £40.60. It is known that there are 9875 items in the total stock.

     (a)   Estimate the total value of the stock to the nearest £10,000.

     (b)   Calculate a 95% confidence interval for the mean value of all items and hence determine a 95% confidence interval for the total value of the stock.

     (c)   If you are a statistician employed by the company, write a brief explanation of the confidence interval that you have obtained that would be easily understood by management.

     (d)   You are told this interval is too wide for decision purposes and you are asked to assess how many more items would need to be sampled to obtain an interval with the same degree of confidence, but with one half the width.

17.2   The sample size is 50.

The sample mean is £320.41.

The sample standard deviation is £40.60.

(a)   The best estimate of the population mean is the sample mean.

We estimate the population mean by £320.41.

Hence we estimate the total (population) value by 9875 x £320.41 :

3.16 million £ to nearest £10,000.

(b)   The standard error of the sample mean is $\frac{40.60}{\sqrt{50}}$ = 5.74.

The 95% confidence interval for the population mean is formed by taking 2 (or 1.96 to be more exact) standard errors above and below the sample mean.

Hence the 95% confidence interval for the mean value for all items in the audit is:

320.41 ± 2 x 5.74

ie (£308.93, £331.89)

The 95% confidence interval for the total value is obtained by multiplying this by the total number of items:

(3.05, 3.28) million £

(c)   We have estimated the total stock value to be 3.16 million £ but this is based on a sample of 50 items and, although we hope that it is representative, we cannot expect it to be exactly equal to the true value of the total stock.

In order to express how close we are with this estimate, we have calculated a 95% confidence interval. This is an interval, centred on our estimate, in which we are confident that the true value of the total stock lies. Then we are able to say that whatever the true value is, we are 95% sure that it lies in the interval (3.05, 3.28) million £.

(d)   Since the 95% confidence interval is two standard errors above and below the sample mean then its width is four standard errors.

The 95% confidence interval for the population mean had a width of 331.89 -308.93 = £22.96, so that half of this is £11.48.

Now, for a sample of n items the standard error of the sample mean is 40.6/√n so that we require four of these to be equal to 11.48.

ie   4 x 40.6/√n = 11.48

giving n = 200.1 .

and since n must be a whole number we need a sample size of 201. We have already sampled 50 so that we shall need to sample a further 151 items.

17.3   Explanatory Notes

1.   There are two main branches of basic statistical inference, namely estimation and hypothesis testing, and these are very closely related.

2.   This question is concerned with the estimation of the average value and total value of the totality of items in the stock from a random sample of 50 items. We use the sample mean to estimate the population mean and multiply the sample mean by the total number of items to estimate the population total. These are known as point estimates and, by and large, point estimates are usually intuitively very obvious. The more challenging and interesting problem is to somehow express the accuracy of these estimates and this is where confidence intervals play their part. Since we don't expect the point estimate to be exactly equal to the population mean we calculate a Confidence Interval; this is an interval, situated about the point estimate, in which we expect to have 'captured' the true mean. If the interval is wide then we are clearly saying that we don't have much confidence in our point estimate as the true mean could be a long way from it. If the interval is short, however, then we expect the true mean to be close to the sample mean and so we have a good estimate. This concept is easily understood by a non-statistician and so provides a useful way of expressing the accuracy of an estimator in layman's terms.

3.   The justification of the confidence interval as being the sample mean $\pm$ two standard errors is easily understood. In the earlier example we studied the sampling distribution of the sample mean and saw that there was a 95% chance of the sample mean falling in the interval,

population mean $\pm$ 1.96 x standard error.

The multiplier, 1.96, is the value in the standard normal table (Example 12) that contains 95% of the probability between $\pm$ 1.96. It is common practice to use 2 instead of 1.96 and we have used this multiplier in the solution. Now we see that the sample mean lies within 2 standard errors of the population mean so that there is a 95% chance that the difference between them is less than 2 standard errors. It follows immediately that the population mean must lie within 2 standard errors of the sample mean thus justifying the confidence interval. It should now be easy to see how to calculate intervals with other levels of confidence; eg for a 90% interval we use 1.65 standard errors since 90% of the standard normal lies within $\pm$ 1.65; for a 99% interval we use 2.58 standard errors; etc.

4.   The interpretation of confidence intervals is made to look rather complicated by many statistical texts. Basically it is simple: the 95% expresses our degree of confidence in the interval containing the true mean and if many intervals are calculated in this fashion, then 19 out of 20 will contain the true mean. If we are gamblers we might say that the odds are 19 to 1 in favour of the interval containing the true mean. Purists might take issue with this interpretation but for all practical intents and purposes this is reasonable and clear.

For convenient reference, a summary of the basic confidence intervals is given in Table 6 at the end of this book.

5.   With the above points in mind the solution to this exercise should be readily understood. In part (a) the obvious estimates are given. In part (b) the 95% confidence interval for the population mean is calculated as 2 standard errors on either side of the sample mean. Here we use the result that the standard error of a sample mean is $1/\sqrt{n}$ x the population standard deviation, where n is the sample size. Note that we use the sample standard deviation in place of the population standard deviation because the latter is unknown.

17.4   Explanatory Notes (continued)

5. (continued)

For small samples we would need to make an adjustment to allow for the fact that we are using an estimate of standard deviation rather than the true value; in large samples, however, this adjustment is negligible and can be ignored.   The small sample case is demonstrated in Example 18.

6.   Part (d) of this question demonstrates how the confidence interval concept can be used to decide on the sample size.   If the user is able to state the level of confidence required and the desired width of the interval then it is possible to determine a unique sample size to achieve this.   The method of doing this is demonstrated in the example.   Note that the sample size is fairly large; this reflects the fact that the management have demanded a fairly accurate estimate, represented in the short length of the confidence interval.   The example demonstrates a general principle on the determination of sample size. A commonly posed question is 'How large a sample size should I take?'.   This can only be answered if the user is able to express the 'quality of inference' that is required as in the estimation problem discussed above. Once the user is able to clearly express his requirement in terms of confidence intervals, error probabilities, etc, then it is possible to determine the sample size which will satisfy these.

Main Points

- The confidence interval is easy to interpret and provides a convenient method of expressing the accuracy of an estimate.

- The 95% confidence interval for the population mean is formed by taking 2 standard errors above and below the sample mean.

- This only applies when either the population standard deviation is known or the sample size is large. By large we mean about 30 or more.

18.1   Let us consider the same sampling situation as in Example 17 but this time only a sample of 12 items is taken.

Assuming these to be randomly sampled and to give a sample mean of £320 and a sample standard deviation of £40, determine a 95% confidence interval for the mean value of all items.

18.2   The sample size is 12.

The sample mean is £320.

The sample standard deviation is £40 and this is based on (12-1) = 11 degrees of freedom.

The estimated standard error of the sample mean is $40/\sqrt{12} = 11.5$ and this is based on 11 degrees of freedom.

Since the sample size is small and we are using an estimate of the standard deviation then we must use Student's t-distribution (Table 4 at the end of this book) and the 95% confidence interval for the population mean is

$$320 \pm t_{11} \times 11.5 \ ,$$

where t_{11} is the value of Student's t that contains 95% of the probability between $\pm t_{11}$.

From the table we see that this is equal to 2.2.

Thus the 95% confidence interval is {£294.7, £345.3}

18.3   Explanatory Notes

1.   The major distinction between this and the previous example is that the confidence interval in this case has to be based on calculations from a small sample. If the population standard deviation were known then the method of Example 17 would still be used even when the sample is small. When we have to estimate the population standard deviation by a sample standard deviation, however, we must take into account the fact that we are in a weaker position due to the sampling variability of a standard deviation computed from a small sample. The adjustment we need to make is to use Student's t-distribution rather than the standard normal when computing the number of (estimated) standard errors above and below the sample mean. This distribution takes into account the fact that we are using an estimated standard deviation rather than the true population standard deviation; it is tabulated in Table 4, and is described below.

2.   The t value that is used depends on the Degrees of Freedom associated with the standard error. (This is really only relevant when the sample size is small). Clearly, the accuracy of an estimated standard deviation depends on the sample size and this is almost the same as degrees of freedom, but not quite! When a sample standard deviation is computed from a single sample then its associated degrees of freedom are one less than the sample size. Degrees of freedom, therefore, can be regarded as 'an adjusted sample size'; there are strong theoretical reasons for this adjustment but we are not able to enter into those details here. It is useful to regard degrees of freedom as an equivalent sample size for the estimation of variability; if we had a sample size of 1, for example, we clearly can say nothing about variablility and we have zero degrees of freedom.

3.   Student's t-distribution : In statistical inference in general this is an extremely important distribution as it takes into account the small sample variability of the sample standard deviation when making inferences about population means. In many statistical accounting problems, however, its application is likely to be limited as sample sizes will often be fairly large and the methods of Example 17 will be more appropriate.

The distribution has a bell shape similar to that of the standard normal but is flatter and more spread out. It is also symmetric about zero. Unlike the standard normal the t-distribution depends on the sample size via the degrees of freedom. In reading Table 4, the first thing to do is to look down the left-hand column to the appropriate degrees of freedom; in this case these are equal to 11. Then we look across the row corresponding to 11 degrees of freedom until we find the appropriate t value. The value at the head of each column represents the area (probability) under the distribution curve and to the right of the t value. Hence for a 95% confidence interval we look under the column with a probability value of $2\frac{1}{2}$% and find the appropriate t value to be 2.201.

The reader should check that, for 11 degrees of freedom:

for a 90% confidence interval the t value is 1.796

for a 99% confidence interval the t value is 3.106 .

These t values are then multiplied by the estimated standard error to compute the required confidence interval.

18.4   Main Points

- When the sample size is small, and the population standard deviation is unknown, then the t-distribution must be used as the multiplier of the estimated standard error in the computation of the confidence interval.

- To find the correct t value from the tables we must know the degrees of freedom associated with the standard error. For a single sample these are simply the sample size less one.

19.1   A random sample of 234 potential voters was taken in three working class areas of Leeds in February 1982.

Of the 111 men sampled, 40 stated their intentions to vote for the Liberal-SDP Alliance in the next election.   Of the 123 women sampled, 27 were in favour of the Alliance. Estimate the difference in the population proportions of men and women who favour the Alliance, and determine a 95% confidence interval for this difference.

Is this strong evidence that men and women differ in their attitude to the Alliance?

Explain the limitations of your estimates with regard to drawing conclusions about the difference in attitudes of the total populations of men and women in Leeds.

[NB: People unwilling to commit themselves to any party are not included in the results of the above survey]

19.2   We estimate the proportion of men in favour of the Alliance by 40/111 = .36.

We estimate the proportion of women in favour of the Alliance by 27/123 = .22.

Then the best estimate of the difference in population proportions is .36 - .22 = .14.
(ie men - women),

The standard error of this estimate is the square root of

$$(.36 \times .64)/111 + (.22 \times .78)/123$$

so that the standard error is .059.

Then the 95% confidence interval for the difference in proportions is

$$.14 \pm 1.96 \times .059$$

ie (.02, .26)   (to 2 decimal places).

We conclude that the percentage of men favouring the Alliance is estimated to be 14%
greater than the corresponding percentage for women, and we are 95% confident that
the true difference lies between 2% and 26%.

Since the confidence interval contains only positive values then we are rejecting 0% as a
possible value of the difference. Hence we are confident that the population proportion
of men favouring the Alliance is greater than that for women.

Our sample is taken from working class areas of Leeds and the estimates, therefore,
apply only to the voting behaviour of working class people. It is quite possible, and likely,
that the difference between men and women would vary betweeen classes and so it would
be wrong to regard this as an estimate for the total population of Leeds.

19.3   Explanatory Notes

1.   This problem deals with the difference in population proportions, and the best estimator for this is quite naturally the corresponding difference in sample proportions. We do not expect this estimate to be exactly right, however, and its accuracy can be conveniently expressed via a confidence interval. As in Examples 17 and 18 we need to know the standard error of our estimate in order to determine the confidence interval.

2.   It can be shown that the standard error of a single sample proportion is the square root of p(1-p)/n where n is the sample size and p is the population proportion. It can also be shown that we get the standard error of the difference of two sample proportions by squaring and adding the individual standard errors and then taking the square root of this sum. Hence the standard error of the difference in proportions is the square root of

    p(1-p)/n + q(1-q)/m.

where p and q are the individual proportions and n and m are the respective sample sizes. Since p and q are unknown we replace them by their estimates so that the standard error is the square root of

    (.36 x .64)/111 + (.22 x.78)/123 .

This is the formula given in Table 6 for confidence intervals.

3.   Now we have to decide how many standard errors are required above and below 14% (the estimate) to give the 95% confidence interval. The appropriate tabulated distribution to use here is the normal, so the relevant multiplier to be used is 1.96; for the normal 95% of the probability is contained in the range -1.96 to +1.96. Note that we use the normal rather than the t-distribution because the sample sizes are larger. In fact we can only use this method of determining confidence intervals for proportions when the sample sizes are large. For small sample sizes the normal is inappropriate and the problem is rather complicated; it is also of little interest as it is impossible to draw any reasonable conclusions about proportions from small samples.

4.   Despite the large sample sizes the confidence interval is quite wide thus showing that accurate estimates for differences in proportions require very large sample sizes. The interval may be interpreted as saying that the true difference in proportions could be anywhere between 2% and 26% but it is much more likely to be near the centre of this interval than the extremes.

5.   It can be inferred from this interval that the proportion of men in favour of the Alliance is greater than the proportion of women in favour; this follows from the fact that 0% is not contained in the confidence interval. Comparing a hypothetical value for a population parameter (in this case zero for the difference in population proportions) is exactly the same as performing a two-sided significance test for that value; see Examples 20 to 22 on this subject, as well as Table 7.

6.   When making an inference from a sample we should be careful to specify the population about which we are drawing conclusions. Since our sample is drawn from the working class population our estimation is for this population and not for the total population of Leeds. This is basically common sense but statistical method is often mis-used in this way.

19.4   Main Points

- Estimates should always be accompanied by a confidence interval in order to express the accuracy of estimation.

- For proportions (and means) the confidence interval is formed by a certain number of standard errors above and below the estimate.

- The multiplier for the standard error is taken from tables of the normal distribution.

- Inferences from sampling only apply to the population from which the sample was drawn.

20.1   A clothes manufacturing company manufactures mens shirts and 10% of these are classified as XL (extra large).

A new type of sports shirt is to be introduced but it is suspected that, for this type of shirt, the demand for XL sizes will not be 10%.

In order to investigate this a pilot study is carried out consisting of the sale of a limited number of the new sports shirts. Of the 200 that sell, 15 are XL.

Carry out an appropriate significance test to assess the strength of this evidence in favour of manufacturing a different proportion of XL sports shirts.

20.2   This is a test for a single proportion.

Let p represent the population proportion of sports shirts sold that will be XL. Then we are interested to know whether this is 10% or not.

Hence we test

$$H_0: p = .10 \quad \underline{v} \quad H_1: p \neq .10$$

We observe a sample proportion of $15/200 = .075$

Under H_0 the standard error of the sample proportion is

$$\sqrt{\left(\frac{.10 \times .90}{200}\right)} = .0212$$

Hence the appropriate test statistic is $(.075 - .10)/.0212 = -1.18$.

Since the alternative hypothesis is two-sided then we carry out a two-tailed test and compare the test statistic with the standard normal distribution. For a test at the 5% significance level the appropriate points are ± 1.96 and since the test statistic is smaller than these (ie more central) then the test is not significant at the 5% level.

The 10% points are ± 1.65 and so we see that the test is not even significant at the 10% level.

We accept the hypothesis that $p = .10$.

On the evidence of these data there is no reason to conclude that the proportion of sales for the new sports shirts that fall in the XL category is not 10%.

20.3   Explanatory Notes

1.   The technique of significance tests (sometimes called hypothesis testing) is used to answer a specific question about a population of interest. A particular statement is made about the population and we wish to decide, in the light of appropriate data, whether or not this statement is true. The statement about the population is known as the Null Hypothesis and often denoted by H_0. The alternative to this statement is known as the Alternative Hypothesis and is represented by H_1.

2.   The first step in carrying out any test of significance is to identify these two hypotheses. This should be done with care as it is often the most difficult and crucial step of the procedure. For this problem we are concerned with a statement about the proportion p of sports shirts sold being classified as XL; we are particularly interested if this is 10% or not and the hypotheses are formulated according to this.

   We now use the data to test the null hypothesis and this is done via a test statistic. There is one appropriate test statistic for each type of significance test and these are shown in Table 7 at the end of this book; once the hypotheses have been identified it is simply a matter of looking up this table and substituting into it to calculate the value of the test statistic. The main objective of a significance test is to measure the compatibility of the observed data with the null hypothesis. If they are found to be incompatible then the null hypothesis is rejected in favour of the alternative hypothesis; otherwise it is accepted. This compatibility is assessed by comparing the test statistic with a standard distribution that it would have if the null hypothesis were true. The appropriate standard distribution is also shown in Table 7. If the value of the test statistic is seen to fall in the 'bulk' of the distribution (ie is one of the more likely values) then it would seem reasonable to conclude that this is its true distribution and that it is therefore compatible with the null hypothesis. If, however, it is seen to be an unlikely value of the distribution then it is incompatible with the null hypothesis and so we reject the null hypothesis.

3.   For this problem the appropriate standard distribution is the standard normal and so we look to see where the value of the test statistic, namely -1.18, falls in the standard normal. It can be seen that this is a reasonably likely value so that these data throw no suspicion on the null hypothesis. This is the crucial concept behind the significance test, namely the decision on compatibility or incompatibility depending on the comparison of the test statistic with the appropriate standard distribution that it should have if the null hypothesis were true. It is logical and intuitively sensible.

4.   It is customary to use Levels of Significance when carrying out a test in order to decide which values will be regarded as compatible and which values as incompatible; that is, where do we draw the line between likely values of the test statistic and unlikely values. Let us suppose that we wish to test at the 5% significance level. Then, for a two tailed test, we look for those points in the standard sampling distribution which cut off tails with probabilities totalling 5% (that is $2\frac{1}{2}$% each). We then regard these low probability values as those showing incompatibility. For the standard normal, for example, the values ±1.96 cut off tails totalling 5% so that if the test statistic falls greater than +1.96 or less than -1.96 then the data are regarded as incompatible with the null hypothesis and we reject it. For the clothes manufacturing problem the test statistic falls well within these limits and so the test is not significant at the 5% level and we accept the null hypothesis that the percentage of XL shirts is 10%. Example 21 contains a further discussion of some practical aspects of levels of significance.

20.4   Explanatory Notes (continued)

5.   The functional form of the test statistic is easily understood. For the basic tests for means and proportions the test statistic is simply the number of standard errors that the sample value deviates frm the hypothetical value; its format, therefore, is (sample value - hypothetical value)/(standard error). This can be seen to be the standardised form of the sample value and should have a standard normal distribution if the hypothetical value (specified by the null hypothesis) is correct.

Main Points

•   A significance test uses data to answer a particular question about a population of interest.

•   The test is based on calculations which assess whether or not the data are compatible with the null hypothesis.

•   The formal tests should be performed in 4 steps:

   (i)   State hypotheses

   (ii)   Calculate test statistic

   (iii)   Compare test statistic with appropriate standard distribution

   (iv)   Draw conclusions in the context of the original (real world) problem.

21.1   A government department wishes to investigate whether the average wage rates of manual labourers in two similar towns, A and B, differ.

Random samples of such workers from the towns gave the following summary statistics

	Average weekly wage	Standard deviation	Sample size
Town A	£95.36	£7.84	120
Town B	£93.02	£9.25	189

It is claimed by the local builders association that the true average wage rates are the same. Use an appropriate significance test to test this claim, and present your answer as a report which can be easily understood by persons with no knowledge of statistical method.

21.2   Report on Wage Rates for Manual Labourers

Purpose of Report

A survey has been conducted in order to assess the validity of the Builders' Association's recent claim that average wage rates in towns A and B are equal. This survey consisted of a random sample of 120 labourers from town A and 180 from town B. In each case the weekly wage of each labourer was recorded. The disparity in sample sizes was based on (a) town B containing a larger number of manual workers, and (b) it is thought that town B wages are more variable than those of town A. For both towns the sample sizes are large enough for the samples to be representative.

Results of survey

The 120 labourers in town A had a mean wage of £95.36.

The 180 labourers in town B had a mean wage of £93.02.

Thus we see that the mean wage of our sample from A exceeds that from B by £2.34.

On the basis of this difference we have to judge the Association's claim. It is important to appreciate that this is a difference in the sample means and, due to sampling variation, sample means will exhibit differences even when the true average wages for all workers are equal. The principal question here is whether this difference is too big to be attributable to sampling fluctuations. Calculations were performed on the salaries of the sampled workers to assess the variability of salaries. The sample mean difference, £2.34, was then compared with the variability; this was done using a standard statistical significance test and this is shown in detail in the appendix below.

Conclusions

The statistical test tells us that there is less than 1 chance in 20 of getting a sample mean difference as large as £2.34 if the Association's claim were true.

There is a clear disagreement between the Association's claim of equal wages and the data resulting from our survey. The average wage in town A is higher than that in town B. This report strongly recommends that we reject the Association's claim. The statistical evidence from our survey gives good support to our action. If a confrontation is envisaged then it may be wise to take a further sample so that the statistical evidence was irrefutable.

Appendix : Statistical Significance Test

Let the population means for all workers be μ_A and μ_B.

Then we wish to use our data to test the hypotheses:

$H_0 : \mu_A = \mu_B$ against $H_1 : \mu_A \neq \mu_B$.

The appropriate test statistic is:

$$\frac{(95.36 - 93.02)}{\sqrt{(\frac{7.84^2}{120} + \frac{9.25^2}{180})}} = 2.35$$

21.3   We now compare this value with the appropriate sampling distribution. In this case it is the standard normal.

Since the alternative hypothesis is two-sided, then a two-tailed test is required. For a test at the 5% significance level we see from the standard normal tables that to reject H_0 the test statistic must be further from zero than 1.96 units.

Hence our test is significant at the 5% level.

Therefore we reject H_0 and conclude that the Association's claim is unjustified.

21.4   <u>Explanatory Notes</u>

1.   The report presents the data and conclusions from the formal test in a form which any intelligent, but non-statistical, reader will understand.   It is important to provide the reader with the underlying logic and reasoning of the procedure but it is equally important to avoid technical jargon and mathematical complexity.

2.   Statistical tests always measure the compatibility of observed data with a hypothesis of interest. If the data are found to be incompatible (ie highly unlikely) with the hypothesis, then the hypothesis is rejected.

3.   The formal layout of the test can be regarded in four stages:

   (a)   State the hypotheses to be tested;

   (b)   Calculate the appropriate test statistic;

   (c)   Compare the test statistic with the appropriate sampling distribution;

   (d)   Draw conclusions from this comparison and state these conclusions in the context of the original (real world) problem.

4.   We first 'translate' the question of interest into a formal statement about hypotheses.   This is often the most difficult part of a test.   In this case we are asking if the population means for the two towns are equal.   The alternative to this is that they are unequal.   Representing the population means by μ_A and μ_B we are interested in whether $\mu_A = \mu_B$ or $\mu_A \neq \mu_B$ and these are our null and alternative hypotheses.   Note that it is conventional to take the simpler of the two hypotheses as the null hypothesis.   It is important to clearly state the alternative hypothesis as the test procedure is related to this.

5.   We next compute a value for the test statistic.   The form of the test statistic is uniquely determined by the hypotheses to be tested; the test statistics for the basic tests are shown in Table 7 at the end of this book.

   In this case it is equal to

   $$(\bar{x}_A - \bar{x}_B)/(\text{standard error})$$

   where the square of the standard error is equal to

   $$(s_A^2/n_A) + (s_B^2/n_B)$$

   and $\bar{x}$, n, s are the sample mean, sample size and sample standard deviation respectively.

   The standard error is the square root of the sum of the squares of the standard errors of $\bar{x}_A$ and $\bar{x}_B$.

   By dividing $(\bar{x}_A - \bar{x}_B)$ by its standard error we obtain a statistic which should have a standard normal distribution if the null hypothesis were true.

6.   The test is then performed by observing where the value of the test statistic falls in the standard normal distribution.   If it falls in the main 'bulk' of the probability then nothing unusual has been observed and the data are compatible with and are supporting the null hypothesis.   If the test statistic falls in a tail, however, then the data is incompatible with the null hypothesis and we reject this hypothesis.

21.5   Explanatory Notes (continued)

7.   Significance levels. These are used to determine which tail regions of the sampling distribution should be used for rejecting the null hypothesis. For a test at the 5% significance level we divide the 5% equally between the tails of the sampling distribution and use cut-off points which cut off tails containing probabilities of .025 (ie $2\frac{1}{2}$%). If the test statistic falls in one of these tails the test is said to be significant at the 5% level.

Other levels of significance can be used. Levels of 10% and 1% are in common use. It should be clear that a test that is significant at a very low significance level provides very strong evidence against the null hypothesis. Conclusions from significance levels may be classified thus

Significance at the 1% level is extremely strong evidence against H_o.

Significance at the 5% level is strong evidence against H_o.

Significance at the 10% level, but not at the 5% level, throws suspicion on the null hypothesis but is not convincingly against it. More data would be desirable so that a firmer conclusion could be drawn.

If a test is significant at the 5% significance level then it means that there is less than one chance in twenty of getting the actually observed data if the null hypothesis were true. This gives a measure of compatibility of data with hypothesis.

8.   For the problem we are considering we apply a 2-tailed test at the 5% significance level. We use a two-tailed test because the alternative hypothesis is two-sided. Hence we look for points in the standard normal distribution which each cuts off a tail with a probability content of $2\frac{1}{2}$%. From the tables we see that these points are at -1.96 and +1.96. The test statistic just falls in the upper tail which implies that the test is significant at the 5% level. Conclusions are then drawn in the context of the original problem.

Main Points

•   This is a test concerned with a comparison of the means of two populations and the hypotheses to be tested must be formulated in terms of these.

•   The compatibility of the data with the null hypothesis can be expressed in terms of the level at which the test is significant.

•   If the test is significant at a very low level (1% say) then the data are very incompatible with the null hypothesis.

22.1   A company is to launch a new make of video recorder onto the market. It is expected that daily sales will average above 100 and if this is so no further advertising will be planned. If early sales indicate that daily sales average below 100 then the present level of advertising will be increased. The first 11 days of sales give the following results:

94, 105, 100, 90, 96, 101, 92, 97, 93, 100, 91.

Carry out a significance test in order to advise the company if more effort should be made to advertise its product.

22.2   Let μ represent the true average daily sales.

Then we wish to test $H_0 : \mu = 100$ v $H_1 : \mu < 100$.

From the data the sample mean = 96.3 and the standard deviation = 4.8 on 10 degrees of freedom.

The the appropriate test statistic = $\dfrac{(96.3 - 100)\sqrt{11}}{4.8}$ = -2.56

We compare this with the t-distribution on 10 degrees of freedom.

For a one-tailed test at the 5% level the appropriate t point is -1.81.

For a one-tailed test at the $2\frac{1}{2}$% level the appropriate t point is -2.23.

The test statistic is more extreme than either of these points.

The test is significant at the $2\frac{1}{2}$% level.

We reject the null hypothesis.

On the evidence of these data the sales are not so high as was expected and a step up in publicity will be required to raise sales to the desired level.

22.3   Explanatory Notes

1. This problem is concerned with the value of the true level of sales. Sales of video recorders will vary from day to day about the true mean and we wish to know if this true mean is below 100; if it is below 100 we will advise an increase in advertising but if it is 100 or more then sales are regarded as satisfactory.

    The first step, therefore, is to set up the formal hypotheses. If we denote the true (unknown) mean sales by μ then we wish to test if μ is 100 against it being less than 100. Note that the alternative hypothesis is, in this case, one sided. An important point to note here is that we do not worry about the possibility of μ being greater than 100. The reason for this is that we will take the same action, namely no advertising, if μ exceeds 100 as when μ is equal to 100. But we will take a different action if μ is less than 100 and this is why we formulate this as the alternative hypothesis.

2. We then observe the data and see that the sample mean of 11 readings is 96.3. The principal question now is whether the deviation of 3.7 below 100 is attributable purely to chance or is it too large to be explained by chance alone. The formal hypothesis test gives the answer to this question and, via the test statistic, it compares this deviation with the measure of its variability, namely the standard error. Since we do not know the population standard deviation we must use its sample estimate and this is computed from the 11 data values in the usual way (see Example 10).

3. The test statistic is then the deviation divided by the standard error giving the value -2.56. As we see from Table 4 we now compare this with the t distribution on 10 degrees of freedom because we are using an estimated standard error. This distribution and the reasons for its use are discussed in detail in Example 18. Bearing in mind the basic logic of the significance test we look to see if the value of the test statistic, namely -2.56, is in the 'bulk' or the tail of the t distribution on 10 degrees of freedom. It looks to be well down in the left-hand tail and this is confirmed when we look at significance levels. For a test at the 5% level we look up, from the tables, the value that cuts off a lower tail with a probability content of .05; note that we put all 5% into <u>one</u> tail because the alternative hypothesis is one sided. We use a one-tailed test.

4. The t point is seen to be -1.81 and since our observed test statistic is more extreme than this then it follows that the test is significant at the 5% level. If, in addition, we look at the $2\frac{1}{2}$% t point this is seen to be -2.23 so that the test is significant at the $2\frac{1}{2}$% level. This is very strong evidence against the null hypothesis and so we conclude that sales are not high enough and more advertising is necessary.

5. It is of interest to understand the correct meaning of significance at the 5% level. It means that if average sales are truly as high as 100 then there is less than 1 chance in 20 of getting observations as extreme as those we have actually experienced. In this case we have significance at the $2\frac{1}{2}$% level so the chance of such observations, if the true mean were 100, is less than 1 in 40. Clearly a case of high incompatibility!

22.4   Main Points

- As we are only interested in detecting deviations of the population mean below 100 then the alternative hypothesis is formulated as a one-sided hypothesis and so we carry out a one-tailed test.

- As we use an estimated standard error rather than the true value then the correct sampling distribution is the t distribution rather than the standard normal.

- Express the level at which a test is significant at the lowest level possible (in this case $2\frac{1}{2}$% rather than 5%). This expresses the confidence of the decision maker that he has made the correct choice.

23.1   The following table shows the numbers of candidates passing, being referred, and failing professional examinations set by three separate bodies A, B and C.

	A	B	C
Pass	233	737	358
Refer	16	68	29
Fail	73	167	136

Test the hypothesis that the chances of pass, refer, fail for the three examining bodies are the same. If any differences are found, indicate briefly where they may be.

23.2   This is a contingency table and we are to test for association between the two attributes, examination performance and examining body.   We do this using a χ^2 test for association.

We test H_0 : no association $\underline{v}$ H_1 : association

The observed values, and the row and column totals, are:

	A	B	C	Total
Pass	233	737	358	1328
Refer	16	68	29	113
Fail	73	167	136	376
Total	322	972	523	1817

For each of the 9 cells we now compute the expected values; each of these is computed as (row total x column total)/grand total.

	A	B	C
Pass	235	710	382
Refer	20	60	33
Fail	67	201	108

The test statistic is

$$\frac{(233 - 235)^2}{235} + \frac{(737 - 710)^2}{710} + \ldots + \frac{(136 - 108)^2}{108}$$

$$= 18.45 .$$

We compare this with the χ^2 distribution on $(3-1) \times (3-1) = 4$ degrees of freedom.
The upper 5% point of this distribution is 9.488.
The upper 1% point of this distribution is 13.28.

This test is significant at the 1% level, and almost significant at the .1% level.
Hence we strongly reject the null hypothesis and conclude that this set of data provides strong evidence of association.

The distribution of results is, therefore, definitely not the same for all three examining bodies.   On comparing the actual frequencies with the expected frequencies it can be seen that:

   They are reasonably close for A.

   For B the observed is greater for pass but smaller for fail.

   For C the observed is smaller for pass but greater for fail.

All three bodies appear to be different from each other, the greatest difference being between B and C. Students do significantly better taking B's examinations than C's.

23.3   Explanatory Notes

1.   This is a standard contingency table for two attributes, examining body and examination result, and the associated test for association between these attributes is a standard test known as a χ^2 (chi-squared, pronounced KI-squared) test. It is called this because the test statistic, which we shall define, has to be compared with a particular tabulated distribution known as the χ^2 distribution.

2.   A contingency table is formed when people (or objects) are classified according to two (or more) attributes; in this case students are classified according to the examinations they take and according to their results in the examinations. Each attribute may have 2 or more levels; in this particular case each has 3 levels. The contingency table, therefore, consists of a set of frequencies which we shall call the observed frequencies.

3.   This may now be used to test for association between the attributes. It is important to understand what association means. There is association when the distribution of frequencies over the levels of one attribute is different for the levels of the other attribute. In the example this would mean that the distribution of results for one examining body was different from that of at least one of the other examining bodies. This association may be interpreted in the other direction: there is association when the distribution of examining bodies for passes is different from the distribution of examining bodies for fails etc. When there is no association the attributes are independent in the sense that the probability of a student passing, failing or being referred is the same for all 3 examining bodies.

4.   The null hypothesis to be tested in this context is no association, and the alternative hypothesis is association; association can occur, of course, in several different ways. The first calculation to be performed is the computation of the table of expected cell frequencies; these are the frequencies we would 'expect' if there were no association and we knew the row and column totals. If we knew only these totals then we would estimate the probability of a student passing by 1328/1817, and we would estimate the probability of a student taking examination A by 322/1817. Hence if the attributes are independent then we can estimate a student passing examination A by

$$\frac{1328}{1817} \times \frac{322}{1817}$$

and since there are 1817 students altogether the expected number in the first cell is

$$1817 \times \frac{1328}{1817} \times \frac{322}{1817} = \frac{1328 \times 322}{1817}$$

This is 235 when rounded to the nearest integer and the expected value is seen to be

(row total x column total)/grand total .

This is the correct calculation for all the other cells and is general in that it applies to all contingency tables.

23.4   Explanatory Notes (continued)

5.   A comparison of the observed and expected frequencies now gives some idea of the compatibility of the data with the null hypothesis. The expected values are calculated on the assumption of the null hypothesis of no association so that if the tables agree then the data are supporting the null hypothesis. Discrepancies between the tables, however, indicate that there is association. There is a simple formal statistic for testing for no association. For each cell we calculate:

$$\frac{(\text{Observed - Expected})^2}{\text{Expected}}$$

and sum these values over all the cells. It can be seen that this will be zero if the tables agree completely and gets larger as the tables disagree more; it can never be negative. Hence large values of this statistic imply association. In the example we are considering the test statistic value is 18.45.

6.   As in any other test of significance we now compare this with a standard sampling distribution in order to make a decision about the hypotheses. The distribution in this case is the χ^2 distribution and this is a tabulated standard distribution; Table 5 at end of book. Like the t distribution it depends on degrees of freedom and these are equal to 4 in the example; the determination of the degrees of freedom is described in the next section. Unlike the normal and t distributions the chi-squared is not symmetric and is for positive values only. The diagram at the top of the Table 5 shows the shape of the distribution. Each row is for a different number of degrees of freedom, the right-hand tail probabilities are shown at the top of each column and these represent the probabilities to the right of the value shown in the body of the table. On 4 degrees of freedom, for example, there is a probability of .10 of χ^2 exceeding 1.533 and a probability of .01 of it exceeding 3.747. Our table shows only the probability tails for the smaller probabilities as these are the only ones we shall use in the examples. A fuller table would show these tails for the whole probability range.

7.   The degrees of freedom for a two-way contingency table are always $(r-1) \times (c-1)$ where r is the number of rows and c is the number of columns. For the example we have a 3 x 3 contingency table and so there are 4 degrees of freedom. This simple computation of the degrees of freedom may be explained thus for an r x c contingency table; we have rc observed frequencies and this is our starting point for determining the degrees of freedom. To determine the expected frequencies we need to know the row totals, column totals, and grand total. It is easily seen that once we know the grand total we need only to know $(r-1)$ of the row totals and $(c-1)$ of the column totals. We lose a degree of freedom for every total that we need to know in order to calculate the expected frequencies. Hence the degrees of freedom associated with χ^2 are

$$rc - 1 - (r-1) - (c-1),$$

and a little basic algebra shows that this is equivalent to $(r-1)(c-1)$.

8.   Once the significance of association has been established we should study the table to see if we can interpret the type of association that is present. Once again it is the cells with the largest discrepancies between oberved and expected values that provide the main clues. These, together with an understanding of the meaning of association, lead to a meaningful interpretation. This approach is demonstrated clearly in the example where it is deduced that examining boards B and C differ appreciably in their results. Note that if the test for association had been found to be 'not significant', then there would be no point in looking for any further interpretation.

23.5   Main Points

- A two-way table of frequencies, where the frequencies are classified according to two attributes, is known as a contingency table.

- A standard χ^2 test may be used to test for association between the attributes.

- Expected frequencies are computed as (row total x column total)/grand total.

- The degrees of freedom = (r-1) x (c-1).

- When association is found to be significant it may be interpreted by comparing observed frequencies with expected frequencies.

- The null hypothesis is always 'no association'.

24.1   Explain the strength and type of relationship you would expect to find between two variables if you are told that their correlation coefficient is:

   (a)   + .94
   (b)   - .02
   (c)   - .97

The number of insurance salesmen employed by a company varies from month to month. These numbers, together with the value of insurance sold, are shown in the following table for a period of 12 months.

No. of salesmen	Monthly sales in £000
215	2206
225	2210
214	2200
238	2218
224	2204
205	2201
231	2215
220	2212
237	2220
210	2204
226	2216
232	2212

Draw a scatter diagram of monthly sales by numbers of salesmen and calculate the correlation coefficient for these data. By referring to the scatter diagram state whether or not you consider the correlation to give a fair representation of the strength of the relationship between insurance sales and numbers of salesmen.

24.2 (a) There is a very strong positive relationship between the variables. A scatter diagram would show the points to almost lie on a straight line with a positive slope.

(b) There is no <u>linear</u> relationship between the variables. This does not necessarily mean, however, that they are not related. There may well be a strong non-linear relationship. A look at a scatter diagram of the points would decide whether the small correlation coefficient was due to the independence of the variables or a relationship that was not linear.

(c) This is the reverse of (a). There is again a very strong linear relationship between the variables but this time the line has a negative slope. This means that if one variable is above average then it is very likely that the other will be below average.

<u>Scatter diagram of sales by number of salesmen</u>

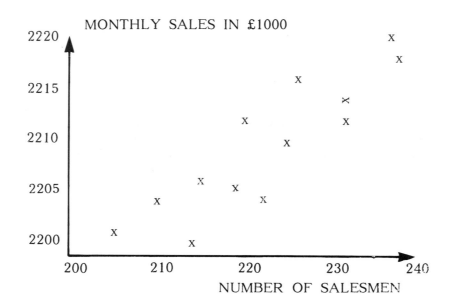

The scatter diagram shows that there is a 'linear shape' to the scatter of the points, but the scatter is fairly wide about this line. The 'trend' is definitely upward, however, indicating a positive relationship between the variables.

24.3   To calculate the correlation coefficient let y = sales; x = number of salesmen. Number
       of data pairs is 12. the calculations are set out as a table:

	x	y
Sum	2677	26,518
Mean	223.08	2209.83
Sum of squares	598,421	58,600,862
Corrected sum of squares	1226.92	501.67
Sample variance	111.54	45.61
Sample S.D.	10.56	6.75

Sum of products	5,916,410
Corrected sum of products	686.17
Sample covariance	62.38

Then the sample correlation coefficient $= \dfrac{\text{covariance}}{(\text{SD of x})\,(\text{SD of y})} = \dfrac{62.38}{10.56 \times 6.75}$

The Sample Correlation Coefficient for insurance sales and numbers of salesmen is .875.

24.4   Explanatory Notes

1.   The first step to be taken in assessing the relationship between two variables is to plot a scatter diagram. This gives an immediate and effective visual assessment of the relationship. The plot of sales by numbers of salesmen shows a definite upward trend with some scatter about it. We are, therefore, sure there is a positive relationship and it appears to be linear; there is no indication of non-linearities.

2.   The formal statistic that is used to measure a relationship is the correlation coefficient. If it is calculated from a data set then it is the Sample Correlation Coefficient; if the data set becomes larger and larger then this statistic eventually 'settles down' to become the population correlation coefficient. It is another statistic to add to the sample means and variances. The sample means and variances measure the individual locations and spreads of two data sets, but the sample correlation measures the strength of the relationship between them. Below we shall discuss some details of its calculation but we begin by discussing its meaning and properties.

3.   It is a mathematical fact that the correlation coefficient lies on a scale ranging from -1 to +1. High positive values imply a strong positive linear relationship between the variables such as in case (a). A correlation close to -1 would also indicate a strong linear relationship but this time it is an inverse relationship so that high x values are associated with low y values and vice versa. Less extreme values of a correlation, and particularly values close to zero, need to be interpreted with caution. It is tempting to say that they imply a weak relationship or independence but this is not necessarily the case as the following two diagrams illustrate. In each case the correlation is near to zero.

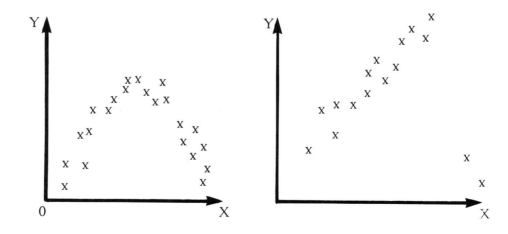

24.5   Explanatory Notes (continued)

4.   In the first case the correlation is zero because there is certainly no linear relationship but there is a clear curved (quadratic) relationship and this can be seen from the scatter diagram.

5.   In the second case the correlation is zero because of the two outliers but it is clear from the scatter diagram that, if these are removed, then there is a very strong positive relationship. In practice we should ask questions about these outliers to ascertain why they are so far out; in most cases a good reason will turn up (such as typing error, mis-reading, etc) and we can discard them with confidence.

6.   The lesson to be learnt from these diagrams is that we should be careful in interpreting small correlations, and a scatter diagram should be consulted before making any firm decisions. If the scatter is seen to be all over the place and with no patterns , then we are justified in inferring that the variables are independent.

7.   We next consider the calculation of the correlation coefficient from paired data. Computers and many hand-calculators will calculate a correlation coefficient and this is clearly the best way to determine its values. It is helpful, however, to see an example of a calculation as this gives confidence in understanding this particular statistic as well as providing a means of determining it when a computer or sophisticated calculator is unavailable. The correlation coefficient is defined to be:

   covariance/(product of the two standard deviations)

8.   The sample covariance may be computed in several equivalent ways. Let us suppose we have n pairs of data and these are denoted by

$$(x_1, y_1) (x_2, y_2) ..... (x_n, y_n).$$

We then 'correct' all the x's by the x mean and all the y's by the y mean giving

$$(x_1' y_1') (x_2' y_2') ...... (x_n', y_n')$$

where $x_1' = (x_1 - x\ mean)$ $y_1' = (y_1 - y\ mean)$ etc.

Then we form $x_1' y_1' + x_2' y_2' + ... + x_n' y_n'$

and on dividing this by (n-1) we arrive at the covariance.

It can now be seen that if a relationship is positive then x_1', y_1' etc. are likely to have the same sign so that their product is positive. If the variables have an inverse relationship then the products will generally be negative. Hence the covariance provides a measure of the strength of the relationship. There is one snag with this statistic. however, since its value depends on the units in which the original data were measured. We overcome this by dividing the covariance by the product of the individual standard deviations of the variables and this final statistic is the correlation coefficient. It does not depend on the units of measurement and always lies on the scale -1 to +1 and the interpretation of its value on this scale has already been discussed in detail.

24.6   Explanatory Notes (continued)

9.   There is, however, an alternative method of calculating the correlation coefficient and it is this method which has been used in the solution to the example.   The calculations are set out in the form of a table the first 6 lines of which are concerned with the calculation of the standard deviations; the reader is referred to Example 10 for a reminder of this calculation.   Line 7 of the tables shows 'the sum of products'; this is sometimes called the 'raw sum of products' and is simply the sum of the values formed by multiplying the original x values by their corresponding y values.

Line 8 gives the corrected sum of products.   We first form a statistic known as the 'correction for the means'; this is the product of the two totals (line 1) divided by the number of data pairs, namely 12.   This 'correction for the means' is then subtracted from the 'raw sum of products' (line 7) to give the value shown in line 8.

We then divide the 'corrected sum of products' by 11, the number of data pairs less 1, to give the sample covariance in line 9.   We finally divide this by the product of the standard deviations to give the correlation coefficient.

10.   In the example we find the correlation coefficient to be .875 thus indicating that there is a strong positive relationship between the variables.   There is a fair scatter about the linear trend but there is no doubt that insurance sales and numbers of salesmen are strongly associated and we could use the known number of salesmen to predict the insurance sales for a month; this is the subject of Examples 26 and 27.

Main Points

- A scatter diagram should always be drawn to assess the strength of a relationship between variables.

- The correlation coefficient provides a useful formal measure of the strength of the relationship.

- Low values of a correlation coefficient should be interpreted in conjunction with the scatter diagram.

- The correlation coefficient lies on a scale from -1 to +1.

- The reader should become familiar with a method of calculating the correlation coefficient.

25.1   Ten examination candidates took an examination in accounting and an examination in statistics. Their marks, out of 100, for each of these papers are shown in the following table.

Calculate the rank correlation coefficient for these marks and comment on the relationship between the marks of the two papers on the basis of this coefficient.

Accountancy	Statistics	Accountancy	Statistics
61	82	51	52
69	63	43	54
73	74	77	72
63	69	46	50
56	62	68	64

25.2   We first determine the rankings of the students for each of the examination papers. These are shown below in brackets next to the original marks.

Accountancy	Statistics	Accountancy	Statistics
61 (6)	82 (1)	51 (8)	52 (9)
69 (3)	63 (6)	43 (10)	54 (8)
73 (2)	74 (2)	77 (1)	72 (3)
63 (5)	69 (4)	46 (9)	50 (10)
56 (7)	62 (7)	68 (4)	64 (5)

We now compute the correlation coefficient for the paired rankings instead of the original exam scores.

	Accountancy	Statistics
Total	55	55
Mean	5.5	5.5
Raw sum of squares	385	385
Corrected sum of squares	82.5	82.5
Sample variance	9.17	9.17

Raw sum of products	362
Corrected sum of products	59.5
Sample covariance	6.61

Rank correlation coefficient = .72 .

The rank correlation of .72 is high thus indicating that there is a strong positive relationship between examination scores on the two papers. Candidates who score higher on the accountancy paper also tend to score higher on the statistics paper.

25.3   Explanatory Notes

1. To calculate, the rank correlation coefficient is no more complicated than the standard correlation coefficient. The principal difference is that instead of using the original data for the two variables in order to calculate the correlation, we perform the calculation on the ranks of the scores. Each value is ranked in the group of mesurements on that variable and the usual correlation calculation is then applied to the ranks. This calculation is demonstrated in the example; once again the coefficient could be obtained by feeding the ranks into an appropriate calculator or computer program.

2. There is some simplification in calculating correlation coefficients for ranks; the sample means and sample variances for the two variables must be equal as they are based on the same group of integers. There are, in fact, formulae for these based on the sum and sum of squares of the first n whole numbers but these are of little advantage. There is also a formula for the rank correlation coefficient and many elementary texts give this. It gives exactly the same value as the calculation described above. In some texts this rank correlation coefficient is called the Spearman Correlation Coefficient after its instigator.

3. There are pros and cons for using the rank correlation as opposed to the usual correlation coefficient. Its main advantage is that it is more robust to outliers and significance tests based on it do not depend on distributional assumptions. Replacing the original data by their ranks, however, tends to play down the strength of the correlation and it will often be underestimated by the rank correlation. For the example the correlation coefficient for the original examination scores is also .72, the same as the rank correlation. There is one distinct advantage of rank correlation: when the data are qualitative (eg excellent, good, fair, etc) rather than quantitative, rank correlation can still be used but the ordinary correlation coefficient does not exist.

4. The rank correlation suffers from the same drawbacks as the standard correlation coefficient when it comes to the interpretation of small correlation values. It should then be interpreted in conjunction with a scatter diagram (of ranks); these points were discussed in more detail in Example 24.

5. There is one important point about the interpretation of correlation that applies equally well to the standard correlation coefficient and the rank correlation coefficient, and this is the implication that there is a causal relationship between the variables. The existence of a correlation does not necessarily imply that a change in one variable will cause a change in the other; variables are often highly correlated due to their dependence on one or more other variables which have not been measured. To understand this let us consider the following two cases:

    I   An athlete's performance in an event will be correlated with the amount of time he spends in training for that event. By training longer he will perform better and if he trains less his performance will deteriorate. Training time has a causal effect on performance in the event.

    II   There will be a strong correlation between ice cream sales and suntan lotion sales at a seaside resort. There is no direct causal effect here, however, since the lotion sales will not be increased if more people eat ice cream or vice versa. Both of these variables are responding to sunshine and temperature which are the causal effects; they show a correlation because they both respond in the same direction to the causal variable.

25.4   Main Points

- If the data are qualitative or if we are worried about outliers or the sensitivity of distributional assumptions, then it is often better to measure the strength of a relationship using the rank correlation coefficient rather than the ordinary correlation coefficient.

- This simply involves performing the usual correlation calculation but on the ranks of the data rather than their original values.

26.1   In order to assess the effect that unemployment may have on crime some statistics were
recorded over a certain period of time for comparable inner areas of 12 cities.   The
figures give the number of serious offences recorded by the police in that period and the
average employment figures for the period.   These are shown in the following table.

Number of Offences	Unemployment
6200	2614
4610	1160
5336	1055
5411	1199
5808	2157
6004	2305
5420	1687
5588	1287
5719	1869
6336	2283
5103	1162
5268	1201

(a)   Draw a scatter diagram for these data plotting crime by unemployment.

(b)   Fit a regression line and draw this line on your scatter plot.

(c)   Comment on the relationship and on its representation by a line.

26.2   (a)   <u>Scatter Diagram</u>

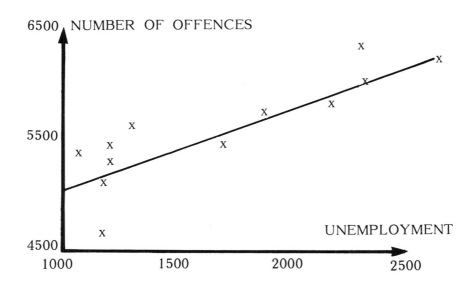

(b)   Calculation of Regression Line

	Unemployment	No. of Offences
Total	19,979	66,803
Sample mean	1664.9	5566.9
Raw sum of squares	36,695,129	374,471,231
Corrected sum of squares	3,431,758.9	2,584,496.9
Sample variance	311,978.08	234,954.27
Sample S.D.	558.55	484.72

Raw sum of products	113,784,494
Corrected sum of products	2,563,065.9
Sample covariance	233,005.99
Sample correlation	.86

Slope of regression line = correlation $\times \dfrac{\text{S.D. for offences}}{\text{S.D. for unemployment}}$ = .75

Intercept = (offence mean) - (unemployment mean) x .75 = 4318.2

Best fitting line is y = 4318.2 + .75 x where x = unemployment and y = no. of offences.

To draw this line on the scatter diagram we need to determine two points on the line and join these.

When x = 1000   y = 5068
When x = 2500   y = 6193

(c)   There is a high positive correlation between crime and unemployment and it can be seen from the scatter diagram that a straight line represents this relationship very well.   There is a fair amount of scatter about the line and one point in particular (1160,4610) appears to be an outlier.   On average it appears that we should expect about 75 more crimes over the length of time period studied for every additional 100 unemployed.

26.3   Explanatory Notes

1.   From the scatter diagram it can be seen that there is a general 'upward drift' of crime with unemployment. The scatter of points is roughly linear but there does appear to be a wider scatter of points at the lower values; the point (1160,4618) is particularly 'off line' and it may be better to leave this out of the analysis. There may well be justification for this if this result is closely examined; for example, the police in that particular inner area may be working to a different definition of serious crime.

2.   It is natural to fit a straight line to such a data set; this could be done by eye but there is a formal method, known as least squares which is widely used in statistics, numerical analysis, etc. and we shall regard the line fitted by this method as the best fitting line. The principle of least squares is fairly easy to understand. If we draw any line through the scatter of points and measure all the deviations of the points from the line in the vertical direction then we can regard these deviations as 'errors'; that is, they are due to random error. It can be shown that there is a unique line which is such that the sum of the squares of these deviations is smaller for this line than for any other line. This is called the least squares line, the best fitting line, or simply the regression line.

3.   If we let x represent distances in the horizontal direction and y distances in the vertical direction then a straight line can be represented by the equation,

$$y = a + bx .$$

The b is known as the slope of the line as it represents the amount that y increases when x increases by 1 unit, and a is known as the intercept of the line as it is the value of y when x = 0. The values a,b, therefore define the line uniquely and a 'little calculus behind the scenes' applied to the least squares principle gives simple formulae for a and b for the regression line. These are

$$b = \text{correlation} \times \frac{\text{S.D. of } y}{\text{S.D. of } x}$$

[**NB**: An alternative formula for the slope is given in Example 27]

and $a = (y \text{ mean}) - b \times (x \text{ mean})$.

4.   It can be seen from the above formulae that the slope of the best fitting line and the correlation coefficient are very clearly associated. In fact they are equal if the variables have the same variability, and their signs are always the same. If two variables are uncorrelated then the regression line will have zero slope. The slope of a regression line therefore measures the strength of (linear) relationship between two variables but, unlike correlation, it is not confined to a finite range.

5.   The variables in a regression calculation are given names. The y-variable (number of offences) is called the dependent variable and the x-variable (unemployment) is called the independent variable.

26.4   Explanatory Notes (continued)

6.   These formulae are used to compute the slope and intercept from the data on crime and unemployment. The resulting line is then drawn on the scatter diagram; to do this we determine two, well separated, points on the line and simply join them. It should be noted that the point formed by the two data means, namely (1664.9, 5566.9) will always lie on the line. The line is a fairly good representation of the relationship between the variables except for the single point corresponding to a low number of offences.

7.   After considering the explanation of least squares in Note 2 above it is natural to ask why we only measure deviations in the vertical direction; why not in the horizontal direction or perpendicular to the line and does it make any difference? The answer to this last question is yes, it does make a difference; measuring deviations in the horizontal direction is equivalent to reversing the roles of the variables and regressing unemployment on number of offences. This would give a different (but similar) regression line. When examining the relationship between two variables we often regard one variable as causing changes in the other or we may wish to use one variable to predict or estimate the other. The regression line we have computed in this example could be used to estimate numbers of offences from unemployment figures for other cities. In such cases the causal variable, or the variable from which we are estimating, should play the role of the independent variable (x) and the responding variable, or variable we wish to estimate, should play the role of the dependent variable (y). It is then correct to consider deviations in the vertical direction only.

8.   The full name for the technique used in this example is Simple Linear Regression. The adjective 'simple' is used to denote that we are explaining the variation in y by only one other variable. There is a generalisation of this technique known as Multiple Linear Regression in which the variation in the dependent variable is related to the values of several independent variables; we do not consider this technique in this text but many texts on statistics and regression give an account of it and almost all statistical computer packages are able to fit multiple regressions.

Main Points

   •   When we examine a scatter diagram it is often appropriate to fit a straight line through the points.

   •   There are simple formulae which give the slope and intercept of the best (least squares) fitting line.

   •   This line may be used to predict or estimate one variable from the other.

27.1   A bicycle manufacturer produces 10 batches of similar bicycles and the costs of production are shown in the following table

Quantity produced (in thousands)
2	3	4	4.5	5	6	7	8	10	14

Production cost (in £1000)
100	107	124	128	130	140	152	160	180	216

(a)   Draw a scatter diagram for these data showing cost against quantity.

(b)   Calculate the best fitting line.

(c)   The manufacturer is planning to produce a batch of 11,000 bicycles. Estimate his production costs and give an approximate 95% confidence interval for the true costs.

(d)   If a new batch of bicycles is to be sold at £60 per cycle, estimate the quantity that would need to be produced to break even.

27.2   (a)   <u>Scatter Diagram</u>

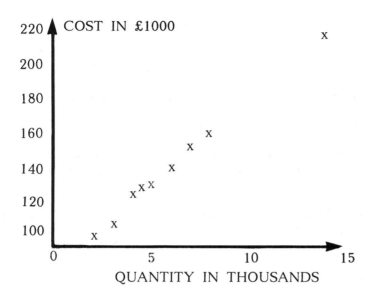

This diagram shows a very linear relationship between the variables. There appears to be a setting up cost of about £85,000 (the intercept) and, after that, the cost per bicycle is constant.

(b)   <u>The Regression Line</u>

The regression line is calculated as in the following table

	Quantity (x)	Cost (y)
Sum	63.5	1437
Mean	6.35	143.7
Raw sum of squares	519.25	217469
Corrected sum of squares	116.025	10972.1
Sample variance	12.89	1219.12

Raw sum of products	10251
Corrected sum of products	1126.05
Sample Covariance	125.12

The slope = $\dfrac{\text{corrected sum of products}}{\text{corrected sum of squares for quantities}} = \dfrac{1126.05}{116.025} = 9.705$

The intercept = (cost mean) - 9.705 x (quantity mean)

= 143.7 - 9.705 x 6.35 = 82.07

The best fitting line is y = 82.07 + 9.705 x.

27.3   (c)   Estimation

To estimate the production costs for 11,000 bicycles we substitute $x = 11$ into the regression equation.
This gives the estimated cost as £188,825.

To measure the accuracy of this estimate we estimate the standard error of the deviations about the line.
This is equal to the square root of

{(corrected sum of squares for y)-(slope)² x (corrected sum of squares for x)}/(n-2)

where n is the number of data pairs.

Then the standard error of the deviations is the square root of

$$\{10972.1 - 9.705^2 \times 116.025\}/8$$

Standard error = 2.35 (in £000)
This is based on 8 degrees of freedom.

Then an approximate 95% confidence interval for the true cost of production is

$$188,825 \pm 2.306 \times 2,350$$

ie (£183,406 , £194,244)

For the production of 11,000 bicycles we estimate the cost at £188,825 and are 95% confident that the true cost will lie between £183,406 and £194,244.

(d)   Break even quantity

If n is the break even number of cycles then the amount received for their sale will be £60n.

From the regression equation the cost of manufacturing n cycles is

£{82,070 + 9.705n}

If the company is to break even then cost = receipts so that

$$82,070 + 9.705n = 60n$$

giving n = 1632 to nearest whole number.

The number of cycles for which the company will break even if each cycle is sold for £60 is estimated to be 1632.

27.4   Explanatory Notes

1. The method used to solve parts (a) and (b) of this example is very similar to that used in the previous example. We have however used a different formula in calculating the slope of the best fitting line. There are often several alternative but equivalent methods of calculating an important statistic; see, for example, the discussion on the calculation of the correlation coefficient in Example 24. The student may well have a favourite method but it is important to be aware that others exist and that they are exactly equivalent in that they will give exactly the same value. Before the advent of powerful calculating aids the choice of method was important as one method would often involve less computational effort than another. This is no longer the case, however, as calculations of correlation coefficients and regression coefficients are easily handled by computers and calculators. Previously we used the formula

$$\text{correlation} \times \left(\frac{\text{S.D. for } y}{\text{S.D. for } x}\right)$$

for the slope coefficient. In this example we have used the alternative formula,

$$\frac{\text{corrected sum of products}}{\text{corrected sum of squares for } x}$$

The student may like to calculate the slope using the first method to verify that they give the same value. Students with algebraic backgrounds should have no difficulty in proving the equivalence of these formulae.

2. From the scatter diagram it can be seen that the points are very close to being linear. The straight line fits these data points very well and the deviations from it are small. We should expect the correlation to be very high and it is, in fact, equal to .9981!

3. Simple linear regression is a model based technique. By this we mean that the statistician views the data as having been generated from a particular mathematical model and this allows him to apply inference techniques such as confidence intervals and significance tests. The model assumes that there exists an underlying true linear relationship between the variables of the form

$$y = A + Bx,$$

but the data we actually observe does not lie exactly on this line due to errors in the observations which may be present for many reasons; they may be experimental errors or deviations due to other variables that we are not able to observe. The full linear regression model may then be written as

$$y = A + Bx + e$$

where e represents the deviation of an observation away from the true line in the y direction. Assumptions are made about the distribution of these error terms; they are assumed to follow the normal distribution, to be centred about zero and to have an error variance σ^2. The constants A, B and σ are the population parameters of the model and, although we may never know their exact values, we may estimate them from the data.

27.5   Explanatory Notes (continued)

4.   The intercept parameter, A, and the slope parameter B, are naturally estimated by the least squares estimates that are used in this and in Example 26. If we are to assess the accuracy of estimates, and in particular the accuracy of any estimates of new values, then we need an estimator of the error variance. This is estimated by the sum of squares of the observed deviations divided by (n-2), where n is the number of data pairs. We denote this by s^2 and it has (n-2) degrees of freedom; this is because we have n data pairs but 2 degrees of freedom are lost due to the estimation of A and B. There is an equivalent and more convenient method of estimating σ^2 and this is by the following formula

$$s^2 = \{(\text{corrected s.s for } y) - (\text{slope})^2 \times (\text{corrected s.s. for } x)\}/(n-2)$$

where s.s. stands for 'sum of squares'. This is the formula used in the example, and it gives the value of 2.35 for s (in £000). The standard error of the deviation is therefore 2.35.

5.   When we consider a new x value (quantity produced) the best estimate of its corresponding production cost is found by substituting this x value into the equation for the regression line. When this is done in the example we arrive at an estimated production cost of £188,825 for 11,000 bicycles. This is only an estimate, however, and we can express its accuracy in a confidence interval. Roughly speaking a 95% confidence interval is given by taking two standard errors above and below the estimate. As we have only 8 degrees of freedom, however, it is more correct to use the t distribution (see discussion in Example 18) and the appropriate t value on 8 degrees of freedom is seen to be 2.306. We finally arrive at the confidence interval by taking 2.306 standard errors on either side of our estimate. This interval is fairly narrow thus indicating that the estimate, based on such a good straight line fit, is very accurate.

6.   In determining this confidence interval we have ignored the fact that the slope and intercept estimates are themselves due to sampling variability. An adjustment may be made to the confidence interval to allow for this but it will be very small for a line fit as good as this. It is also possible to express the accuracy of the slope and intercept estimates via confidence intervals. These are beyond the scope of this text but may be found in any specialised statistical text.

7.   The final section of this example is concerned with a 'break even' quantity of bicycles when bicycles are sold at £60 apiece. To determine this we simply equate the return, as a function of the number of cycles, to the cost given by the regression equation. This is a simple linear equation and can be solved to give the required number of cycles.

Main Points

*   Linear regression is a model based technique and assumes the data are generated by deviations about a true line.

*   The fitted line may be used to give an estimate of the value of the dependent variable corresponding to a new value of the independent variable.

*   The accuracy of this estimate may be expressed by a confidence interval centred on the estimate; to calculate the confidence interval it is necessary to estimate the standard error of the deviations about the line. There is a standard formula for this.

28.1   The following time series represents the index numbers for a company's quarterly turnover from 1982 to 1985. These are based on an index of 100 for the first quarter of 1982.

		1982	1983	1984	1985
Quarter	1	100	109	118	128
	2	103	112	121	-
	3	108	116	126	-
	4	129	140	153	-

(a)   Compute an appropriate moving average for this time series and plot it, together with the original series on a graph.

(b)   Assuming an additive model, compute seasonal adjustments for the four quarters.

(c)   Determine forecasts of the indices for the remaining three quarters of 1985.

(d)   If the company's turnover in the first quarter of 1982 was 2 million pounds, translate your forecasts of the indices into forecasts of actual turnover.

28.2 (a)   We compute a 4 point moving average to eliminate seasonal effect and then average consecutive terms in this to centre it.   The final column shows the deviation of the index from the trend.

	Time Period	Index	4 point Moving average	Central Moving Average (trend)	Deviation
1982	1	100			
	2	103	110.0		
	3	108	112.3	111.2	-3.2
	4	129	114.5	113.4	+15.6
1983	1	109	116.5	115.5	-6.5
	2	112	119.3	117.9	-5.9
	3	116	121.5	120.4	-4.4
	4	140	123.8	122.7	+17.4
1984	1	118	126.3	125.1	-7.1
	2	121	129.5	127.9	-6.9
	3	126	132.0	130.8	-4.8
	4	153			
1985	1	128			

Diagram

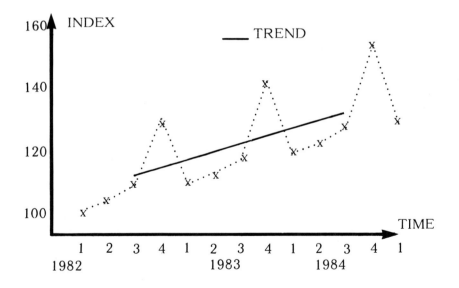

From the diagram we see that the moving average is almost completely linear and so we have represented it by a straight line.   The series of indices deviates appreciably from this trend and it is clear that there is a strong seasonal effect on the company's turnover.

28.3 (b)   To compute seasonal adjustments we average the deviations of index from trend for each of the four quarters and then adjust these averages to have an average of zero.

Quarter	Average of deviations	Seasonal adjustment
1	-6.8	-6.6
2	-6.4	-6.2
3	-4.1	-3.9
4	+16.5	+16.7

(c)   Since the trend is linear the average increase per quarter in the trend value is

$$\frac{130.8-111.2}{8} = 2.45$$

Using this and the seasonal adjustments the index forecasts for the remaining three quarters of 1985 are (to nearest whole number)

                                    Forecast

1985	2	130.8 + 2.45 x 3 - 6.2 = 132
	3	130.8 + 2.45 x 4 - 3.9 = 138
	4	130.8 + 2.45 x 5 + 16.7 = 160

(d)   To convert these index forecasts into forecasts of the company's turnover we note that the index of 100 corresponds to two million pounds of turnover.

Hence we need to multiply the indices by a factor of 2,000,000/100 = 20,000.

Turnover Forecasts (in £million)

1985 2nd quarter	2.64
1985 3rd quarter	2.76
1985 4th quarter	3.20

28.4 Explanatory Notes

1.   Index numbers are often used to represent time series because of their simplicity and the ease with which we can compare them with a simple base (100 in this case). If the original series involves values to several decimal places or very large numbers then we can get a better grasp of the variability of the series by working with index numbers. In this case the first quarter of 1982 is used as a base of 100 when the company's turnover for this quarter was two million pounds. To convert turnover to index numbers we have multiplied by a factor of 100/2,000,000 and to reverse the operation, as we need to do in section (d), we simply multiply by 2,000,000/100 = 20,000.

2.   The additive model assumed for this series is of the form

     Series = Trend + Seasonal Component + Error

     There is little we can do about the error component which should average to zero.

     The seasonal component varies over the four quarters and these four seasonal adjustments are assumed to sum to zero. Then if we average any four consecutive terms in the series the seasonal component is eliminated and we are left with trend. This is done in column 3 of the table in part (a) ; this shows the averages for each set of 4 consecutive terms of the series. Note that each of these contains each of the 4 quarters. If the original series had been monthly readings then we would have used a 12 point moving average at this stage. The term, 'moving average' for these averages is self-explanatory. Unfortunately the 'centre points' of these moving averages do not coincide with the quarters and so they cannot be directly compared with the original series. We overcome this by averaging each two consecutive terms of the 4 point moving average series thus forming our final moving average.

     This is centred on the quarters and so we can compare this directly with the original series. The deviations of the series from this trend are given in the final column of the table in section (a); this is simply column 2 minus column 4. The trend is in fact a weighted 5 point moving average; this is discussed in detail in Example 29. Note that the trend series is shorter than the original series as we are not able to compute trend values at the extremes of the series.

3.   The diagram shows time plots of the original series and the trend values. The trend is seen to be extremely linear and we also see that large positive deviations of the series above trend occur for each of the fourth quarters; these indicate a strong seasonal effect.

4.   We use the deviations of series from trend to estimate the seasonal components. These computations are shown in the table in section (b). The deviations for the 3rd quarter are -3.2, -4.4, -4.8 and these are averaged in column 2 to give the effect of the third quarter. The sum of the averages of deviations for the quarters is not zero and so we adjust them by their average so that the final seasonal adjustments (column 3) sum to zero. To see how this is done it can be seen that the average of column 2 is -0.2; then by adding 0.2 to each value in column 2 we arrive at the seasonal adjustments in column 3. A similar method would be applied to a monthly time series and in that case we would have 12 seasonal adjustments.

28.5   Explanatory Notes (continued)

5.   The seasonal adjustments show that the company's turnover is always greatest for the fourth quarter, and smallest for the first two quarters of the year.

6.   We use the additive model to make forecasts. Each forecast is composed of the extrapolated trend plus the appropriate seasonal adjustment. Extrapolating the trend can sometimes be difficult (see Example 29) but in this example it is easy as it is so obviously linear; we have to have a certain amount of faith in that it will continue that way! If certain economic occurences suggest otherwise then we should do our best to change it according to our subjective judgment.

In this case the trend can be extrapolated by continuing the line on the graph or by calculation; we see that the trend increases by 2.45 units per quarter. Since our first forecast (2nd quarter 1985) is three quarters after our last trend value of 130.8 then we must increase 130.8 by three trend increments to estimate the trend value for the 2nd quarter of 1985. Hence the estimated trend value is 130.8 + 3 x 2.45 = 138.15. This is then adjusted by the seasonal adjustment for the second quarter to give the final forecast. The forecast of the index for the other two quarters of 1985 are calculated in a similar manner, and these are converted into forecasts of the company's turnover as we have already described in note 1.

Main Points

- Index numbers are simply a scaled version of the original series and are easier to assess and compare.

- A time series can often be regarded as being composed of a trend plus a seasonal component.

- The trend is estimated by an appropriate moving average.

- The seasonal component is estimated from the deviations of the series from the trend.

- Forecasts are based on the extrapolated trend plus the seasonal adjustment.

29.1   The following table shows the average price (in each quarter) of new dwellings from 1978 to 1981.

Year	Quarter	Average Price
1978	1	19,575
	2	20,700
	3	22,230
	4	23,640
1979	1	24,915
	2	26,730
	3	28,590
	4	30,270
1980	1	31,665
	2	32,895
	3	33,510
	4	33,765
1981	1	34,305
	2	34,440
	3	33,675

Calculate a set of index numbers to represent these average prices using 1978 as a base of 100.

For these indices represent the trend by a suitable moving average and plot these indices and the trend by time. Use this trend to compute seasonal adjustments for each of the four quarters. Construct a forecast of the index for the fourth quarter of 1981 and calculate the corresponding forecast of average dwelling price.

29.2   To arrive at an index value of 100 for 1978 (1) we have to multiply the average dwelling price for that quarter by 100/19,575. Multiplying all the other prices by this factor and rounding off to the first decimal place gives the following set of indices:

Year	Quarter	Index	Year	Quarter	Index
1978	1	100	1980	1	161.8
	2	105.7		2	168.0
	3	113.6		3	171.2
	4	120.8		4	172.5
1979	1	127.3	1981	1	175.2
	2	136.6		2	175.9
	3	146.1		3	172.0
	4	154.6			

We represent trend by a 5 point moving average with weights equal to

$$1/8\{1, 2, 2, 2, 1\}$$

This gives the following trend values, centred on the quarters

Year	Quarter	Index	Year	Quarter	Index
1978	1	-	1980	1	160.8
	2	-		2	166.1
	3	113.4		3	170.1
	4	120.7		4	172.7
1979	1	128.6	1981	1	173.8
	2	136.9		2	-
	3	145.5		3	-
	4	153.7			

Diagram of Index and Trend

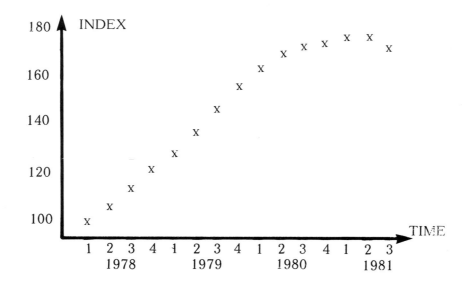

29.3   From the diagram it is clear that the basic indices show a very smooth trend so that the deviations from the moving average are small. The trend is linear until the end of 1979 and it then falls away; it is still rising but not nearly so steeply as in 1978 and 1979.

To compute the seasonal adjustments we compute the deviation of the indices from their corresponding trend values for each quarter and average these within quarters. These are adjusted to sum to zero.

Quarter	Deviation	Average	Seasonal Adjustment
1	-1.3, 1.0, 1.4	+0.37	-.26
2	-0.3, 1.9, 2.1	+1.23	+.61
3	0.2, 0.6, 1.1	+0.63	+.01
4	0.1, 0.9, -0.2	+0.27	-.36

Due to the smoothness of the series of indices these seasonal adjustments are very small. There is clearly very little seasonal effect in dwelling prices.

Since the trend is clearly non-linear we cannot compute a simple increment per quarter as we did in Example 28. Extrapolation of such a trend is much more difficult to achieve by technical means and a simple extrapolation by eye, using the graph, is probably as good a method as any.

From the graph we see that the extrapolated trend for the index number is about 175 for the fourth quarter of 1981. Applying the seasonal adjustment of -.36 we arrive at an index forecast of 174.64.

Converting this back into an average dwelling price we reverse the previous operation and multiply by the factor 19,575/100. This gives the forecast of the average dwelling price for the fourth quarter of 1981 as £34,186 to the nearest £.

29.4   Explanatory Notes

1.   The method of computing index numbers, the trend and seasonal adjustments for this example is exactly the same as the method of the previous example and needs no further comment except for the trend computation which, at first sight, appears to be different.   It is, however, algebraically equivalent and if the method of Example 28 were used then exactly the same values would result.   To see this let us represent the time series as

$$x_1, x_2, x_3, x_4, x_5, x_6, \ldots$$

Then, using the method of trend calculation of Example 28, we first compute 4 point moving averages

$$\frac{1}{4}(x_1 + x_2 + x_3 + x_4)$$

and

$$\frac{1}{4}(x_2 + x_3 + x_4 + x_5).$$

and then 'centre' these by averaging each two consecutive terms to give the trend representation as

$$\frac{1}{2}\left\{\frac{1}{4}(x_1 + x_2 + x_3 + x_4) + \frac{1}{4}(x_2 + x_3 + x_4 + x_5)\right\}$$

and this is equal to $\frac{1}{8}\{x_1 + 2x_2 + 2x_3 + 2x_4 + x_5\}.$

This is the weighted 5 point moving average used in this example.

2.   The resulting calculations from this example differ from those in Example 28 in two respects.   The original series is itself very smooth and is very close to the trend so that there is very little seasonal variation.   The second difference is that the trend is non-linear which makes extrapolation of the trend for forecasting purposes much more difficult.   In this example we suggest that a 'by eye' extrapolation using the graph is sufficient.   Forecasting, as desirable and necessary as it may be, is often an exercise in 'crystal ball gazing' and we must accept that it will contain a subjective element.

3.   The models used to analyse the data in the example and in Example 28 were additive models.   It is also possible to use a multiplication model of the form

   series = trend x seasonal component

This may be more appropriate when we expect the seasonal deviation to increase as the values in the series get larger.   This can be analysed in various ways.   We could, for example, assume the seasonal components to multiply to unity and represent the trend as geometric moving averages.   Alternatively we could take logs and thus reduce the multiplicative model to an additive model and analyse it accordingly.   The student should be aware of these other possibilities but should concentrate his attention on the analysis of the additive model.

4.   When we subtract trend and seasonal adjustment from a series we are left with error.   This may not be completely random, however, and the series that remains is known as a stationary time series.   There is a vast literature on this subject and we mention it here only in case the student is baffled by the high powered mathematics that may be seen in books on time series.   This further analysis is very much a case of great effort for diminishing return and the simple 'bread and butter' analysis described in these two examples is sufficient.

29.5   Main Points

- A moving average trend can be computed in two equivalent ways; as an average of averages or as a 5 point weighted moving average.

- When trends are non-linear it is often necessary to extrapolate them by graphical methods rather than by calculation.

30.1   A paper manufacturing company needs to purchase a number of paper cutting machines. There are two possible types of machine and each requires two operators.

One machine costs £8,000 and requires 30 square metres of floor space; this machine cuts 300 sheets of paper per minute.

The other machine costs £15,000 and requires 40 square metres of floor space; this machine cuts 500 sheets of paper per minute.

The total floor space available is 700 square feet and the company has budgeted £200,000 for purchasing the machine. The company has 24 operators in its employment and it is essential that these are fully employed (ie the company must employ at least 24 operators).

(a)   If x represents the number of smaller machines bought and y the number of larger machines bought, express the above constraints as linear inequalities in x and y.

(b)   Draw a graph of these constraint lines and clearly identify the feasible region.

(c)   Determine, subject to the constraints, the numbers of machines which will maximise the production rate.

(d)   Compute the cost of buying these machines and the resulting production rate.

30.2   (a)   x = number of smaller machines
y = number of larger machines

Constraints are :

Cost $8x + 15y < 200$
Space $3x + 4y < 70$
Operators $x + y > 12$

(b)

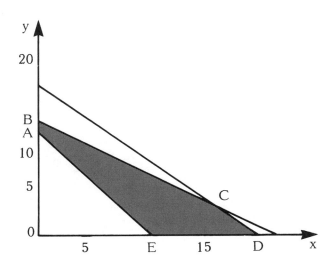

(c)   The production rate, in hundreds of sheets per minute, is

$$P = 3x + 5y$$

This has a slope of -3/5 which is steeper than -8/15 (the cost constraint slope but less steep than -3/4 (the space constraint slope).

For each value of P the production rate can be represented by a line on the graph and to maximise P we require the line with the largest P value which also has a point in the feasible region. Taking into account the slope comparisons above this can be seen to be the line through the point C.

The point C, therefore, represents the optimal feasible point. To find its coordinates we solve the simultaneous equations

$$3x + 4y = 70 \quad (1)$$
$$8x + 15y = 200 \quad (2)$$

(1) x8 gives $24x + 32y = 560$
(2) x3 gives $24x + 45y = 600$

and on subtracting $13y = 40$ so that $y = 3.08$
and on substituting this value back into (1) $x = 19.23$

Hence the values of x and y that maximise P are 3.08 and 19.23 respectively. For machines these have to be whole numbers and so the number of machines the company should buy are 19 small machines and 3 large machines.

It is easily verified that the point (19,3) lies within the feasible region.

30.3   (d)   The cost of buying 19 small machines and 3 large machines is

£8,000 x 19 + £15,000 x 3 = £197,000 .

[**NB:** This is £3,000 within the budget]

The production rate using these machines is

300 x 19 + 500 x 3 = 7,200 sheets per hour .

30.4   Explanatory Notes

1.   Linear programming should not be confused with computer programming; it is the name given to a technique for maximising (or minimising) a linear function of variables which are subject to several linear constraints.

2.   Many examination questions on this topic are 2 variable problems. For 2 variables we can represent the problem graphically using an x,y coordinate system and this is the approach used to solve this particular problem. When there are more than 2 variables the graphical approach is inappropriate and a computational algorithm known as the <u>Simplex Method</u> can be applied. This is described in Example 32. There are, of course, many standard computer programs which will solve a linear programming problem.

3.   There are different types of linear programming problems. The most common type is the 'allocation problem' in which we have to allocate values to variables subject to restraints on those variables. Another commonly arising problem is the 'transportation problem'; this is demonstrated in Example 33.

4.   The first step in a linear programming problem is to construct equations representing the constraints and a linear equation representing the function to be maximised (or minimised). This latter function is known as the <u>objective function</u> and in this case it represents production and so we wish to maximise it. The constraint equations are given in the solution and are self-evident.

5.   In section (b) we represent the constraints on a graph. Let us consider the constraint for space, $3x + 4y < 70$. On the graph we draw the line $3x + 4y = 70$ and this may be regarded as a boundary line between feasible and infeasible points. It is easily seen that points above this line violate the constraint and so are infeasible whereas points below the line satisfy the constraint and so are feasible as far as this particular constraint is concerned. This is repeated for the other two constraints so that we draw the boundary lines,

$$8x + 15y = 200 \quad \text{and} \quad x + y = 12.$$

Note that points above the line $x + y = 12$ are the feasible points. We also have the two natural boundary lines $x = 0$ and $y = 0$ since the number of machines purchased cannot be negative! We now identify the region of points which are feasible with respect to all of the boundary lines, and this is the shaded area shown in the diagram. It is bounded by the figure ABCDE which is called 'a simplex'. We now know that whatever x,y values we choose, they must lie within this feasible region and our objective is to determine the point on this region that has a higher productivity than any other point in the feasible region.

6.   For section (c) we set up the linear function of x and y that represents producttion (in hundreds of sheets per minute). This is $P = 3x + 5y$. Now it can be proved that the optimum point in a problem such as this must lie on one of the vertices of the simplex; ie it must be at one of the intersection points A, B, C, D or E. A simple approach would be to evaluate P at each of these points and then choose the point for which P is largest.

An alternative approach is to visually imagine $P = 3x + 5y$ as a variable line crossing the graph as we increase P. For a fixed value of P this represents a line with a negative slope of $-3/5$ and as we increase P this line 'moves' parallel to itself but in an upward direction to the right. The productivity, P, will be the maximum at the point where this 'moving' line leaves the feasible region; this is easily seen to be C. If we increase P any further when $P = 3x + 5y$ passes through C, this line will fall completely outside the feasible region.

30.5   Explanatory Notes (continued)

7.   We find the point C by solving the simultaneous equations of the two lines that intersect at C. This gives the point C as (19.23, 3.08). Normally this would be the optimum solution for the variables but in this case the solution must consist of whole numbers and so we round off to get $x = 19$ and $y = 3$. A little care is needed in doing this to check (a) that the rounded off solution lies in the feasible region and (b) that there is not a 'nearby' pair of integers that give a larger P value. This complication does not arise when x and y are continuous variables.

8.   Having obtained the solution for the number of machines to be purchased it is an easy matter to compute the corresponding cost and productivity. We simply substitute $x = 19$ and $y = 3$ into the relevant cost and productivity equations.

9.   From this solution it should be clear that a minimisation problem can be treated in a similar manner, but in this case we regard the line representing the objective function as moving from upper right to lower left on the graph. The feasibility region is unaltered. In this example let us assume the constraints are the same as before so that we have the same feasible region in the graph on page 30.2. Let us consider the problem of choosing the numbers of machines to minimise their running costs when we know that the small machine costs £20 a day to run and the large machine costs £30 a day.

Then the daily running cost is $20x + 30y = C$ with a slope of 2/3. If we imagine this line moving across the feasible region, but maintaining its slope, we see that it 'leaves' the feasible region at the point E and so this is the optimal point. In this case we buy no large machines and as few (12) small machines as possible. This is an obvious solution in this case but it serves to demonstrate the graphical approach to a minimising problem.

Main Points

•   Linear programming is a technique which can be used to solve problems involving the maximising (or minimising) of a linear function of variables (eg profit) subject to linear constraints on those variables.

•   With only 2 variables a graphical approach is appropriate.

•   The feasible region is represented by a figure known as a simplex.

•   The optimum point is a vertex of this figure; this is easily identified from the objective function and the solution found by solving the simultaneous equations representing the constraint lines meeting at the vertex.

31.1   A chemical manufacturing company produces two blends of fuel, X and Y. Each fuel is composed of 3 chemicals A, B and C.   The amounts of these chemicals required to manufacture 1 gallon of X and 1 gallon of Y are shown in the following table

	X	Y
A (gallons)	.3	1.1
B (gallons)	1	.2
C (kgms)	1	1

For a single production run the chemical plant cannot intake more than 88 gallons of A, 60 gallons of B and 100 kilograms of C. The nett profit on one gallon of fuel X is £5 and on one gallon of fuel Y is £3.

(a)   Determine the amounts of X and Y that should be manufactured in each production run in order to maximise profit.

(b)   Discuss the incentives to modify the plant in order to intake greater quantities of the 3 chemicals.

31.2   (a)   If the plant produces x gallons of fuel X and y gallons of fuel Y then we wish to maximise the profit

$$P = 5x + 3y$$

subject to the constraints,

$$3x + 11y < 880$$
$$5x + y < 300$$
$$x + y < 100$$

The graph of these constraints is shown below.

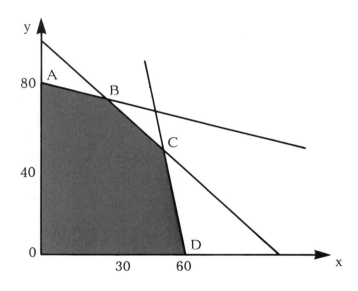

The feasible region is within the figure ABCDE and is shown shaded.

The slope (-5/3) is steeper than that of BC (-1) but not so steep as that of CD (-5). It follows from this that the optimum point is C and we determine its coordinates by solving the following simultaneous equations.

$$x + y = 100$$
$$5x + y = 300$$

giving   x = 50 and y = 50

The plant should therefore produce 50 gallons of each fuel.

For each production run the input to the plant should be

700 gallons of A
300 gallons of B
100 kilograms of C

The nett profit per run would be £400.

31.3   (b)   There is clearly no point in modifying the plant to intake more of constituent A as the full amount of A is not required for optimum production.

Optimum production requires the maximum quantities of B and C, however, and we now assess the gains to be made in modifying the plant to intake greater quantities of these chemicals.

To do this we construct the dual problem to the above linear programming problem and solve this to obtain the shadow prices.

If b and c are the shadow prices for chemicals B and C respectively then from the dual problem we solve the equations,

$$5b + c = 5$$
$$b + c = 3$$

so that b = 0.5 and c = 2.5.

We may then report that:

For every gallon of B above the present limit of 300 gallons we would make an additional profit of £0.5.

For every kilogram of C above the present limit of 100 kilos we would make an additional profit of £2.5.

There would therefore appear to be a greater incentive to modify the plant to intake greater amounts of C than to take in more of B. There is no incentive to intake more of A.

31.4   Explanatory Notes

1.   The first part of this problem is a straightforward linear programming problem for 2 variables and is solved by the graphical method in a similar fashion to Example 30.

2.   In part (b) of this problem we investigate the incentives to relax constraints by computing the Shadow Prices corresponding to these constraints. Only the critical constraints are of interest; for chemical A we only need to use 700 gallons so that there is a spare capacity of 180 gallons. Hence there is no incentive to relax the constraint on chemical A. Chemicals B and C are used at their maximum values, however, and the optimum solution will be improved if more of these chemicals may be used. The increase in profit for every extra gallon of chemical B that may be used is known as the Shadow Price for chemical B. The shadow price for chemical C is the increase in profit that would result if the plant could intake 101 kilograms of C rather than 100.

3.   The most obvious way to compute the shadow price for chemical B is to compute the profit that would result if an intake of 301 gallons of B were used; the increase of this above the profit of £400 is then the shadow price for B.

ie we solve $5x + y = 301$ and $x + y = 100$

giving $x = \dfrac{201}{4}$ $y = \dfrac{199}{4}$

with a profit of £400.5.

Hence the shadow price for B is £0.5.

The student should verify that a similar calculation for chemical C gives a shadow price of £2.5.

4.   There is an alternative method of determining these shadow prices which is simpler and neater; it is the method used in the solution to this example and has the advantage of determining the two shadow prices simultaneously.

We first write down the equations for the boundary lines at the optimum point C together with the objective function. The shadow prices b and c are written alongside the corresponding equations.

Chemical B	$5x + y$	$= 880$	b
Chemical C	$x + y$	$= 100$	c
Profit	$5x + 3y$	$= P$	

We then construct a set of equations for the Dual Problem. For every linear programming problem there exists a dual problem whose solution contains aspects relevant to the main problm. These equations are formed by taking the coefficients in a vertical direction and using them as coefficients in simultaneous equations for b and c thus,

$$5b + c = 5$$
$$b + c = 3$$

On solving these we arrive at the shadow prices $b = £0.5$ and $c = £2.5$.

31.5   Main Points

- Decisions on the value of investment to relax a constraint in a linear programming problem may be judged on the basis of shadow prices.

- Shadow prices only apply to the critical constraints.

- The shadow price corresponding to a constraint is the increase in profit that results from a unit increase in the constraining value.

- Shadow prices may be computed by a direct method or, more easily, by use of the dual problem.

32.1   Compare the graphical method and simplex method for solving linear programming problems.

Demonstrate the simplex approach by applying it to the fuel blends problem of Example 31. Explain clearly your interpretation of the simplex tableau corresponding to the optimal solution.

32.2   The graphical method for solving a linear programming problem is only appropriate to a 2 variable problem because of its dependence on a graphical representation in two dimensions. For 2 variables it has the advantage of simplicity and of being easy to interpret. The simplex method can be applied to any number of variables as it is a purely computational method.   It also has the advantage that the shadow prices are automatically generated by the calculations that determine the optimal solution. It is easily programmed (for a computer) so that solutions may be easily and conveniently obtained.

For the linear programming problem of Example 31 the constraints are

$$3x + 11y < 880 \qquad \text{chemical A}$$
$$5x + y < 300 \qquad \text{chemical B}$$
$$x + y < 100 \qquad \text{chemical C,}$$

and the objective function to be maximised is $P = 5x + 3y$.

The initial simplex tableau is

Tableau 1

$\underline{x}$	$\underline{y}$	$\underline{a}$	$\underline{b}$	$\underline{c}$	Value
3	11	1	0	0	880
5	1	0	1	0	300
1	1	0	0	1	100
5	3	0	0	0	0

where a,b,c are the slack variables corresponding to the constraints on chemicals A,B,C respectively.

This tableau corresponds to a basis a = 880, b = 300, c = 100.

We pivot about the value 5 (shown in bold) to obtain the second tableau. This pivot replaces b by x in the basis.

Tableau 2

$\underline{x}$	$\underline{y}$	$\underline{a}$	$\underline{b}$	$\underline{c}$	Value
0	10.4	1	-0.6	0	700
1	0.2	0	0.2	0	60
0	**0.8**	0	-0.2	1	40
0	2	0	-1	0	-300

This corresponds to a basis a = 700, x = 60, c = 40 with a profit value of 300.

It is not optimal since there is a positive value in the last row.

We pivot about the value 0.8 (shown in bold) to arrive at the third tableau. This pivot replaces c by y in the basis.

32.3   Tableau 3

x	y	a	b	c	Value
0	0	1	2	-13	180
1	0	0	.25	-.25	50
0	1	0	-.25	1.25	50
0	0	0	-0.5	-2.5	-400

This corresponds to a basis a = 180, x = 50, y = 50, with a profit value of 400.

Since all the values in the final row are negatives this is the optimal solution.

The interpretation of this final tableau is as follows:

The 3 zeros in the final row show that the basis is x, y, a so that the slack variables b and c are equal to zero. These zero values mean that chemicals B and C are used completely.

The first row says that a = 180 (since b and c are zero) so that there are 180 gallons of chemical A that are not used.

The second row says that x = 50 and the third row that y = 50.
These are the optimal amounts of fuel to be produced.

The fourth value in the values column shows that the profit corresponding to the optimal solution is £400.

The values in the fourth row of columns b and c are (ignoring negative signs) £0.5 and £2.5 respectively. These are the shadow prices.

For every extra unit of chemical B made available the profit would increase by £0.5.

For every extra unit of chemical C made available the profit would increase by £2.5.

32.4   Explanatory Notes

1.   This example demonstrates the application of the simplex method in a simple context and the interpretation of a simplex tableau. The solution obtained is exactly the same as that derived using the graphical method in Example 31.

2.   The simplex method is based on the introduction of new variables which are used to make the inequality constraints into equalities. One such variable is introduced for each constraint and so we have represented the 3 variables in the example by a, b, c to correspond to the constraints on the 3 chemicals. These variables are called the Slack variables, and they cannot be negative. With the inclusion of these variables the constraint equations and objective function may be written as

$$
\begin{aligned}
3x + 11y \;\; + a \qquad\qquad\quad &= 880 \\
5x + y \qquad\quad + b \qquad\quad &= 300 \\
x + y \qquad\qquad\quad + c \;\; &= 100 \\
5x + 3y \qquad\qquad\qquad\quad &= P
\end{aligned}
$$

Tableau 1 on page 32.2 is simply a way of writing these equations. The coefficients in the table are multiplied by the variable at the head of the column and added to give the value on the right-hand side.

3.   Each tableau corresponds to a feasible solution at a vertex of the simplex diagram on page 31.2. Each pivotal iteration corresponds to a move from one vertex of the simplex to another (better) vertex. At each vertex only 3 of the 5 variables are non-zero. The 3 non-zero variables are called the Basis and each vertex corresponds to a different basis.

4.   We begin the simplex procedure with a simple feasible solution corresponding to the origin in the diagram on page 31.2. At the origin the two main variables are zero and so the basis is composed of the slack variables. If x and y are zero then a = 880, b = 300, c = 100 and this corresponds to a zero profit as no chemicals are used and no fuels produced.

5.   We shall explain the technicalities of the iteration later (Explanatory Note 32.7) but concentrate at present on interpretation. The first iteration takes us from Tableau 1 to Tableau 2. This corresponds to moving from the origin to the vertex D in the simplex diagram on page 31.2. At vertex D chemical B is completely utilised and we produce 60 gallons of fuel X. This can be seen in Tableau 2. The zeros in the last row of the x, a, c columns indicate that this is the basis with values x = 60, a = 700, c = 40 and the profit is £300. This is not an optimal solution as there is still a positive value in the final row.

6.   The final iteration takes us to the optimal solution represented by Tableau 3; this corresponds to vertex C in the simplex diagram. This tableau is interpreted in the solution.

7.   We now describe the pivotal procedure for the iteration from Tableau 1 to Tableau 2. To choose the pivot we select the column with the largest value in the final row. This is the x column.

We then divide the coefficients in this column into the corresponding numbers in the values column to give 293.3, 60, 100 respectively. As 60 is the smallest and thus corresponds to the 5 in the x column this is our pivotal position. This pivoting operation will introduce x into the basis and remove the slack variable b since this has a 1 in the pivotal row.

32.5   <u>Explanatory Notes</u> (continued)

8.   We next divide the pivotal row by 5 giving

        1   .2   0   .2   0 60.

We then subtract multiples of this row from the other rows so that all the other values in the pivotal column become zero. For example we subtract 3 x pivotal row from the first row to give

        0   10.4   1   -0.6   0   700

This is also done for the other rows and we then have Tableau 2. These iterations continue until all the values in the final row of a tableau are negative. This is then the optimal solution.

9.   The simplex procedure may seem rather computationally involved at first but it is easily mastered after a little practice. Its main value is that it can be applied to any number of variables. Certain complications may arise but the method is easily adapted to cope with these; for example we may wish to minimise rather than maximise or we may have constraints of an opposite nature so that we cannot commence with a feasible solution at the origin (see the simplex diagram on page 30.2). These can all be overcome by simple adjustments.

<u>Main Points</u>

- When we have more than 2 variables the graphical method cannot be used and the simplex method is then appropriate to solve a linear programming problem.

- This involves setting up, in a tableau format, a simple feasible solution and then iterating by pivotal operations to the optimal solution.

- The final tableau also contains the shadow prices.

33 LINEAR PROGRAMMING - THE TRANSPORTATION PROBLEM

33.1 An oil company wishes to distribute 2,500,000 gallons of a high grade petrol from 3 refineries to 4 distribution stations. The petrol is transported from a refinery to a distribution site by tankers each of which can carry 10,000 gallons. The costs for a single trip from refinery to site are shown in the following table (in £),

Distribution Site

		1	2	3	4
Refinery	A	20	95	60	15
	B	12	60	24	50
	C	27	33	15	33

The capacity of the site and availability at the refineries are (in 10^4 gallons)

Site	1	2	3	4
Capacity	50	50	100	40

Refinery	A	B	C
Availability	80	60	110

(a) Use the transportation method of linear programming to determine the optimal distribution of petrol in order to completely fill site capacity.

(b) Report on the transportation cost of this procedure.

(c) Which refinery(ies) has/have excess petrol and how much excess does it/they have?

33.2   Since the availability exceeds the capacity we introduce a 'Dummy Site' with a capacity of 250-240 = 10 (in units of 10^4 gallons).

Petrol attributed to this site actually remains at the refinery and the transportation costs to this site are zero for all 3 refineries.

We first construct a feasible solution by allocating tankers to routes in such a way that the allocation is consistent with availability and capacity.   To get a good feasible solution we allocate, as far as is possible, the cheaper routes first.   That is we start with '1 to B' and then '4 to A' and '3 to C' etc. always checking that we have not violated the constraints.   [**NB:** the 'dummy' site is allocated last].

The initial feasible tableau following this approach is:

Tableau 1.1

	1	2	3	4	D	Total
A		30		40	10	80
B	50	10				60
C		10	100			110
Total	50	50	100	40	10	250

We next compute the shadow costs, the marginal 'supply and demand costs' being shown in the margins.   These are:

Tableau 1.2

	1	2	3	4	5	'Supply Cost'
A	47	95	77	15	0	0
B	12	60	42	-20	-35	-35
C	-15	33	15	-47	-62	-62
'Demand cost'	47	95	77	15	0	

By comparing Tableau 1.2 with the transportation costs we see that there are several shadow costs greater than the transport costs.   This solution is, therefore, not optimal and we pivot about the value with the greatest excess (of shadow over transport) to improve the solution.   This is the route 'A to 1' where the excess is £27.

For this pivot the change points are (using Tableau 1.1)

A to 1	+30
A to 2	-30
B to 1	-30
B to 2	+30

33.3   We now consider the new solution

Tableau 2.1

	1	2	3	4	D	Total
A	30			40	10	80
B	20	40				60
C		10	100			110
Total	50	50	100	40	10	250

The shadow costs for this solution are

Tableau 2.2

	1	2	3	4	D	'Supply Cost'
A	20	52	34	15	0	0
B	12	60	42	23	8	8
C	1	33	15	-4	-19	-19
'Demand cost'	20	52	34	15	0	

Comparing this tableau with the transport costs we see that this solution is not optimal, the largest excess (shadow above transport) being 8 for 'B to 3' and 'B to D'.   We therefore use 'B to 3' as pivot, the changes being

B to 3	+40
B to 2	-40
C to 2	+40
C to 3	-40

This gives the solution

Tableau 3.1

	1	2	3	4	D	Total
A	30			40	10	80
B	20		40			60
C		50	60			110
Total	50	50	100	40	10	250

33.4   The corresponding shadow costs are

Tableau 3.2

	1	2	3	4	D	'Supply Cost'
A	20	50	32	15	0	0
B	12	42	24	7	-8	-8
C	3	33	15	-2	-17	-17
'Demand cost'	20	50	32	15	0	

No shadow cost exceeds a transport cost and so this is the optimal solution.

(a)   Tableau 3.1 shows how many tankers should be used on the routes from the 3 refineries to the 4 distribution points.

(b)   The cost (minimised) for this procedure is obtained by multiplying the number of tankers on a route by the cost for the tanker on the route, and summing over all used routes.

This gives $30 \times 20 + 40 \times 15 + 20 \times 12 + \ldots + 60 \times 15 = £4950$

(c)   From Tableau 3.1 we see that there are 10 tankers allocated to the dummy destination from refinery A. This means that 100,000 gallons of petrol refined at A is excess and will not be transported to a distribution point.

33.5   Explanatory Notes

1.   The transportation problem can be expressed as the problem of minimising a linear
     cost function subject to linear constraints and it is, therefore, a linear
     programming problem. Due to its special nature, however, we do not approach it in
     the same way as the allocation problem.

2.   The format of the problem consists of a set of sources (refineries) and a set of
     destinations (distribution points). Each source has a certain quantity of a resource
     that it can supply and each destination has a specified quantity of the resource that
     it requires. The total demand need not equal the total supply.

     Associated with each route (and source-destination combination) is the cost for
     transporting a single unit of the resource from the source to the destination.

     The solution to a transportation linear programming problem is a specification of
     the allocation of resources to routes which minimise the cost of transportation
     subject to the constraints on demand and supply.

3.   These costs and constraints are clearly specified in the problem but we see that the
     supply exceeds the demand by 100,000 gallons. The procedure for solving this
     problem requires demand and supply to be equal; we easily achieve this by
     introducing a 'dummy' source with a demand of 100,000 gallons thus balancing
     demand and supply. Being allocated to this 'dummy' source really means that the
     resource does not leave the supply point; routes between supply points and 'dummy'
     are therefore given a zero transportation cost. A similar device could be used if
     demand exceeded supply; in this case we would insert a dummy supply row.

4.   The general procedure is similar to that of the simplex method described in
     Example 32. We begin by constructing a feasible (but usually non-optimal) solution
     and then carry out a set of iterations which improve the solution at each iterative
     step until we finally arrive at a feasible solution which we cannot improve - the
     optimal solution. We now describe these steps in detail.

5.   It is an easy matter to construct an initial feasible solution as this simply means
     finding any allocation that satisfies the constraints. One method, for example, is
     the 'North West Corner' method; this involves allocating as large a quantity as the
     constraints will allow to the A1 cell and then working across the row until the row
     constraint is satisfied. We then work down the column until that column constraint
     is satisfied, then across the row, and so on until all constraints are satisfied and a
     feasible solution is obtained. Note that the construction of a feasible solution does
     not involve the costs as it is aimed purely at finding any solution that satisfies the
     constraints. The student may like to verify that the application of the North West
     Corner method to this problem yields the following initial feasible solution

	1	2	3	4	D	Total
	50	30				80
		20	40			60
			60	40	10	110
Total	50	50	100	40	10	250

33.6   Explanatory Notes (continued)

6.   Although any initial feasible solution will suffice as a starting position, the number of iterations and computational labour will be reduced by a good choice of the initial solution.   In the solution to this problem we have taken the costs into account in constructing the initial feasible solution.   In this case we allocate the resources in the order of costs; we allocate as much as we can to the cheapest route, then to the next to cheapest, etc.   We follow the procedure as much as we can, subject to the constraints.   In the example the routes may be ordered thus in terms of cheapest upwards [**NB**: we do not include the 'dummy' routes in this list].

B1, A4, C3, A1, B3, C1, C2

We now allocate as much as we can to B1; we allocate 50 units as this is the column constraint.

We then allocate 40 units to A4; this satisfies the column constraint.

We then allocate 100 units to C3; this satisfies the column constraint.

We cannot allocate anything to A1 as column 1 is already satisfied.

We cannot allocate anything to B3 as column 3 is already satisfied.

We cannot allocate anything to C1 as column 1 is already satisfied.

We allocate 10 units to C2; this satisfies the row constraint.
etc.

This gives Tableau 1.1 and we now apply the first step of the iteration to this solution.

7.   Each step of the iteration involves the computation of a set of <u>Shadow Costs</u> which can be used to improve the solution; when the solution cannot be improved it is optimal.   The calculation of these shadow costs may be described in 3 stages.

Stage 1: Set up a cost tableau but insert the transportation costs only in cells which have resources allocated to them thus

	1	2	3	4	D
A		95		15	10
B	12	60			
C		33	15		

Stage 2: Construct pseudo source and distribution costs. These are costs associated with the refineries and distribution points which sum to the costs in the above table.   As a reference value we take the first row available (refinery A) to have a zero cost.   Then distribution points 2, 4 and D must have costs 95, 15 and 10 respectively.   Then refinery B must have a cost of -35 so that this plus 95 gives the B2 cell value of 60.   This is continued until we have costs associated with all 3 refineries and all 5 distribution points.   These are shown as the marginal costs in Tableau 1.2 on page 33.2.

Stage 3: Tableau 1.2 is then completed by inserting, in each empty cell, the sum of the corresponding row and column costs. This is the tableau of shadow costs.

33.7   Explanatory Notes (continued)

8.   We next compare the shadow costs with the transportation costs; ie we compare Tableau 1.2 with the tableau of transportation costs given in the question on page 33.1. If no shadow cost exceeds a transportation cost then the solution is optimal and we have minimised the cost. If there is at least one shadow cost greater than a transportation cost then an improvement can be made and we need to proceed to the next step of the iteration. In the example there are several larger shadow costs and so we proceed.

9.   To improve the solution, when this is possible, we identify the cell (route) for which the excess of shadow cost over transport cost is maximum. In the example this is seen to be for cell A1 where the excess is £27. We then use cell A1 as a pivot to arrive at a better feasible solution. The pivotal operation proceeds as follows.

We intend to move resources into A1 and so A1 is given a + sign. To balance row A we need to remove some resource from a used cell in this row and so we allocate a – sign to A2. To balance column 2 we need to add some resource to a used cell in this column and so we allocate a + sign to B2. To balance row B we need to subtract some resource from a used cell in this row and so we allocate a – sign to cel B1. This loss from column 1 is balanced by the allocation to A1, and so we have now completed the circuit. We now change these cells, according to these signs, by the maximum possible which is 30 units in this case. A1 and B2 increase by 30 and A2 and B1 decrease by 30 so that feasibility is retained and it is easily shown that the overall cost is reduced. After this pivotal operation the new feasible solution is shown in Tableau 2.1 which is the starting point for the next iteration.

The cost for Tableau 1.1 is £6480. The cost for tableau 2.1 is £5670 so that the pivotal operation has reduced the cost by £810. This is easily shown to be equal to (amount of resource moved) x (shadow cost – transport cost) = 30 x £27 = £810.

10.   This procedure is then repeated until we finally arrive at a step in which no shadow cost exceeds a transport cost. The allocation is then optimal and this occurs on the 3rd iteration for this problem. With a bad initial feasible solution it could take much longer.

The minimised cost associated with the final solution is obtained by multiplying the source allocation for the chosen cells by the transport cost for that cell and summing these over the cells. This comes to £4950 which is a reduction of £1530 below the initial solution. Incidentally the North West Corner initial solution given in explanatory note 5 has an associated cost of £8230 which is quite a bit more than the initial solution that we have used; it is likely that several iterations would have been needed if we had started with the North West Corner method.

The amount of petrol not transported is, of course, 100,000 gallons which is the excess of supply over demand; it is the 'dummy' column which tells us what happens to this. In this problem it is seen to remain at refinery A since cell AD has an allocation of 10 units.

11.   The number of used cells in a feasible solution must be equal to

number of rows + number of columns – 1 .

Most of the time this will arise naturally (as it does in the example) and there will be no problems. Sometimes, however, a feasible solution involves less than this number and the solution is said to be degenerate. The result of this is that

33.8   Explanatory Notes (continued)

11. (continued)

problems will arise when calculating the marginal costs as in stage 2 of explanatory note 7, and in pivoting. This is overcome by simply treating some unused cells as used cells to make the used cells up to the correct number; ie they are treated as used cells with an allocation of zero. Care must be taken in choosing such cells and they should be chosen in positions to make the shadow cost calculation and pivoting operation possible. The best method is to choose them when actually calculating the marginal costs at stage 2. When the calculation 'runs out' a cell can be chosen to allow it to continue.

12.   The transportation linear programming technique works perfectly well for a minimising operation. We simply reverse our objective. We choose cells with the largest profit and look for shadow profits less than transport profits, etc. Alternatively we could change the maximising problem into one of minimising by applying negative signs to all the profits.

Main Points

- The transportation linear programming technique applies to problems of allocating resources to routes between sources and destinations when the costs of the routes are known.

- The technique consists of constructing a feasible initial solution and then iterating to an optimal solution using shadow costs.

- The number of iterations will be reduced if we take care to begin with a good initial solution.

34.1   A company of marketing consultants undertakes to survey consumer behaviour in the purchasing of dairy products. The Chairman of the project committee proposes the following breakdown of the project into stages together with their interdependencies and the time and manpower required to complete each stage.

	Activity	Dependence	Duration	Manpower
A	Pilot study	–	2	3
B	General computer planning	–	1	2
C	Analysis of pilot study	A	1	1
D	Questionnaire design	C	1	2
E	Main survey	D	4	4
F	Data collection	E	2	2
G	Final specification of computer requirements	D	1	1
H	Computer programming	B,D	2	2
I	Program testing	H	2	2
J	Computer run on main data	F,G,I	1	1
K	Analysis of output	J	1	2
L	Final report	K	2	3

(a)   Construct a network diagram for this project. Identify the critical path activities and give the duration time of the project.

(b)   Calculate the total, free and independent floats for the activities.

(c)   The company may use its own employees together with personnel from an outside bureau. A person hired from the outside bureau costs the company £800 per man-week. If a company employee is assigned to the project then his assignment is for the whole duration of the project at a cost to the company of £4000.

Assuming that the project must be completed in as short a time as possible determine the most economic scheduling of the activities and determine the optimum staffing.

34.2   (a)   Network Diagram

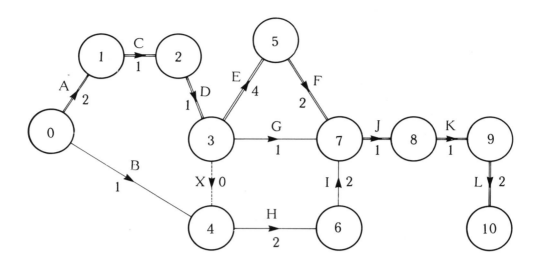

The total time associated with the critical path is 14 weeks.
The critical path is composed of the activities A C D E F J K L.

(b)   The following table shows the floats for the activities.   The formulae used to
      compute these were

            {EST = earliest starting time ; LST = latest starting time }

      Total Float   = LST - EST

      Free Float   = (EST of following activity) - (sum of EST and duration)

      Independent Float   = (EST of following activity) - (duration) -
                             - (sum of LST and duration of preceeding activity).

Activity		Connecting Nodes	Duration	EST	LST	Total	Free	Independent
A	*	0,1	2	0	0	0	0	0
B		0,4	1	0	5	5	3	3
C	*	1,2	1	2	2	0	0	0
D	*	2,3	1	3	3	0	0	0
E	*	3,5	4	4	4	0	0	0
F	*	5,7	2	8	8	0	0	0
G		3,7	1	4	9	5	5	5
H		4,6	2	4	6	2	0	-2
I		6,7	2	6	8	2	2	0
J	*	7,8	1	10	10	0	0	0
K	*	8,9	1	11	11	0	0	0
L	*	9,10	2	12	12	0	0	0
X		3,4	0	4	6	2	0	0

* = critical activities.

34.3 (c) We first consider the scheduling of the critical activities over the 14 week period. This can be represented in a block diagram thus

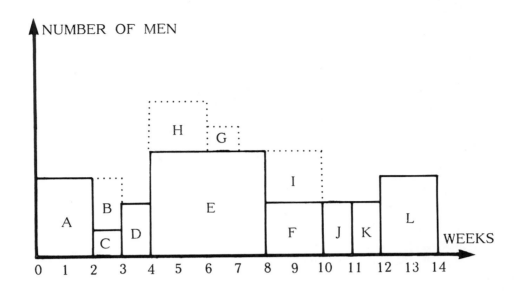

The remaining activities have floats and are allocated to time periods in the order of 'tightness' on their floats and subject to their time order in the network diagram. Their allocation is shown by the dotted boxes in the diagram. The order in which they were allocated was H, I, B, G.

We now cost the possibilities of using 0, 1, 2, 3, 4, 5, 6 company employees, the remaining labour being allocated to bureau personnel.

We note that the total labour involved in this project is 49 man-weeks.
The costing comparison is shown in the following table

No. of Company Employees	Man weeks Covered	No. of Bureau Personnel	Cost
0	0	49	£39,200
1	14	35	£32,000
2	27	22	£25,600
3	38	11	£20,800
4	44	5	£20,000
5	47	2	£21,600
6	49	0	£24,000

From this we see that it is most economic to use 4 company employees and 5 man-weeks from the bureau.

34.4   (c)   (continued)

The schedule for the project would be as follows

Weeks	Schedule
1 + 2	3 company employees on activity A
3	2 employess on B and 1 employee of C
4	2 employees on D
5 + 6	2 employees on H, 2 employees and 2 personnel on E
7	1 employee on G, 3 employees and 1 personnel on E
9 + 10	2 employees on I and 2 employees on F
11	1 employee on J
12	2 employees on K
13 + 14	3 employees on L .

34.5   Explanatory Notes

1.   The technique of critical path analysis provides a systematic method for scheduling the stages (activities) of a project in a time efficient way. The output from this technique is a diagram and a set of meaningful statistics which give the user a clear picture of the project, pinpoint the activities that need most attention, and determine the allocation of manpower to the activities.

2.   The information required to conduct a critical path analysis is shown in the table in the question on page 34.1. It consists of a set of project activities, their dependence on earlier activities and their duration time. (The manpower information is not needed until later). This information is conveniently represented by the network diagram on page 34.2.

3.   This diagram consists of a set of lines, representing the activities, and nodes (circles) where one activity takes over from another. The nodes are numbered for reference purposes and the time direction of the activities is indicated by the arrows on the lines. A node at the beginning of an activity is known as its 'tail' node, and the node at the end is known as its 'head' node. For example, 4 is the tail node of H and 6 is its head node. This diagram gives a very effective visual interpretation of the time dependence of the activity.

An activity cannot be started until the activities entering its 'tail' node have been completed.

An activity has to be completed before all activities leaving its 'head' node can be started.

The diagram is drawn beginning on the left-hand side and moving right across the page. We begin with activities A and B which have no predecessors; then C follows A, D follows C, etc. It will often take a couple of rough sketches to get the right diagram and the final diagram can easily be checked by seeing that the immediate dependencies at each node correspond to the dependencies in the original information table.

4.   It is sometimes necessary to include one or more 'dummy' activities in the diagram to avoid ambiguities. Let us suppose we had set up the diagram incorrectly and had merged nodes 3 and 4 so that the activities associated with the node were of the form

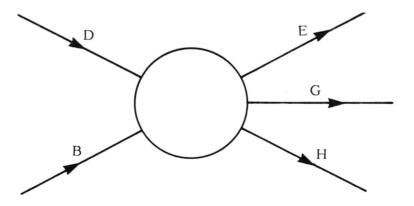

34.6   Explanatory Notes (continued)

4.   (continued)

Checking this against the information table we see the following

H depends on the completion of B and D which is correct.

E appears to depend on B and D which is wrong as E only depends on the completion of D.

G appears to depend on B and D which is wrong as G only depends on the completion of D.

The check shows us that the diagram is incorrect and the way to represent H as dependent on B and D but E and G only dependent on D is via the dummy activity X as shown in the main diagram. This dummy makes H dependent on D but avoids representing E and G as being dependent on D.

5.   We are now in a position to identify the critical path on the diagram. On the diagram there is a number associated with each activity; this is its duration. A duration of zero is allocated to the dummy activity X. We identify all the continuous paths through the diagram, always following the directions of the arrows, and sum the durations for these paths. These give

        A C D G J K L   9 weeks
        A C D E F J K L   14 weeks
        A C D X H I J K L   12 weeks
        B H I J K L   9 weeks

The critical path is the longest duration path, namely A C D E F J K L. The activities on this path are the critical activities. If each of these keeps to time it should be possible to complete the project in 14 weeks but if any of these take longer than specified then the project will be delayed. It is important, therefore, that these activities are kept to their target times. The other paths have a certain amount of float; these are seen to be 5 weeks, 2 weeks and 5 weeks. There are known as total floats. A path with a small float is near to being critical and its activities should also be monitored carefully. The critical path is denoted by a double line in the diagram.

6.   The concept of float time can be analysed into 3 types of float and this is done in the table on page 34.3.

We list the activities together with their 'head' and 'tail' nodes as these are needed in the subsequent calculations.

The Earliest Starting Time for an activity is the time value, measured from the project starting point, at which the activity can begin at the very earliest; it is impossible for it to commence before this time point. We calculate the EST by summing the durations of all the activities on paths leading to the 'tail' node of the activity in question. If there is more than one path then the EST is the maximum of these duration totals. For example, for the paths leading to node 7 the total duration times are 10 weeks and 5 weeks, the EST for activity J is therefore 10 weeks.

34.7   Explanatory Notes (continued)

6.   (continued)

The Latest Starting Times are computed in a similar manner but using paths to the end point of the project. For an activity we determine the path from its head node to the end that has the largest total duration. This is added to the activity's duration and the sum subtracted from 14, the duration of the critical path. For example, from node 3 there are 3 paths to the end with durations of 10 weeks, 5 weeks and 8 weeks. The LST for activity D is therefore 14 - (10+1) = 3 weeks.

It is easily seen that the ESTs and LSTs for critical activities will be equal but this will not be so for non-critical activities.

7.   The Total Float for an activity is the maximum time that the activity can be delayed without increasing the overall duration time for the project. Since the earliest and latest times it can be started without affecting are the EST and LST respectively it follows that

Total Float = LST - EST .

8.   The Free Float for an activity is the maximum time an activity can be delayed without affecting the earliest starting times of following activities. It can be seen from this definition that if the EST of the following activity exceeds the EST of the activity by more than the duration of the activity then there will be a free float. This excess is, in fact, the value of the free float so that we have

Free Float = EST (Following activity) - EST - duration .

For example if we consider activity G in the table on page 34.3, we see that the head node for G is 7 and this is the tail node for J; J is the following activity. We subtract the sum of the EST and duration of G from the EST of J to give

10 - (4+1) = 5 weeks,

and this is the free float for G.

9.   The Independent Float for an activity is the maximum time an activity can be delayed without having any possible effect on the preceeding and following activities. It can be seen from this definition that such a float will exist if the time difference between the latest finishing time (= latest starting time + duration) of the preceeding activity and the earliest starting time of the following activity exceeds the duration time of the activity of interest. This excess, if it exists, is the independent float so that this is equal to

EST (following) - (duration) - (sum of LST and duration of preceeding) .

For example, for activity G we see from its tail and head nodes (3,7) that the preceeding and following activities are D and J respectively. It follows that its independent float is equal to 10 - 1 - (3+1) = 5 weeks.

With such a large independent float it is clear that there is great flexibility for scheduling the activity G. It can be started any time after 4 weeks and before 10 weeks without affecting the scheduling of any other activity.

10.   The final section of this question is concerned with the allocation of manpower to the activities. The solution to this is largely self-explanatory. The technique involves the manipulation of the schedule to reduce the cost of manpower and the network diagram, critical path, and floats are extremely useful in this manipulation.

34.8   Explanatory Notes (continued)

10.   (continued)

We first construct a block diagram of men $\underline{v}$ time for the critical activities as shown on page 34.3. This part of the schedule is inflexible. We then insert the non-critical activities in such a way as to minimise the maximum manpower required. The floats are useful guides at this point and we insert the activities in order of their tightness on floats. That is, the least flexible (smallest floats) first. The positions of these activities in the schedule is shown in the block diagram.

11.   It now remains to cost this schedule for the possible distributions of labour between company employees and bureau personnel. This costing is fully described in the solution and it is seen that the use of 4 company employees results in the minimum cost for the schedule.

Main Points

- A project involving interdependent activities is usefully represented as a network diagram.

- The critical path draws attention to the activities which need careful monitoring to ensure that they do not exceed their target times.

- Other paths can be seen to be near critical or not according to their total floats.

- Three types of float can be computed. These are useful in assessing the flexibility of these activites in the schedule.

35.1   The basic formula used to calculate the economic order quantity (EOQ) for stock is

$$EOQ = \sqrt{\left(\frac{2dc}{h}\right)},$$

where   d = demand per unit period
         c = cost of an order
         h = holding cost per unit period

(a)   Describe the basic reorder level system to which this applies and derive this formula.

(b)   Describe how you would use this formula in the case of variable demand and variable lead time.

(c)   You are to advise a retailer on the inventory control, using a reorder system, for a particular brand of dishwasher. The monthly demand for these dishwashers follows a Poisson distribution with a mean of 6. The lead time for an order from the manufacturer is 2 months and the total costs associated with each order come to £150. Each dishwasher is valued at £200 and the stockholding costs are 20% per annum of stock value.

Determine the EOQ, reorder level and buffer stock for this system if the chance of running out of stock must be less than 5%.

35.2 (a)   We assume a model where demand is at a constant rate, d per unit period.

The number of orders placed per unit period is d/x, and the cost of these is dc/x.

The average amount of stock held is x/2, and the cost of this holding per unit time is xh/2.

Hence the total cost per unit time is cd/x + xh/2, and we choose x to minimise this cost.

Differentiating with respect to x and equating to zero gives

$$- \frac{dc}{x^2} + \frac{h}{2} = 0$$

so that $x = \sqrt{(\frac{2dc}{h})}$, and this is the EOQ.

It is easily verified that this gives the minimum of the cost function.

(b)   If demand is constant and there is a lead time t for new orders then the EOQ is given by the same formula but the order would have to be placed at a time t before stock ran out; ie when the stock level has fallen to td. This is the reorder level. For variable demand and lead times the formula can be applied with the demand per unit time replaced by expected demand per unit time.

The variable lead time does not affect the EOQ but the order would be placed at a time before the expiry of stock equal to the expected lead time. A buffer stock would also have to be introduced to cope with larger than average lead times or large demands occuring in the lead time.

(c)   Working to a unit time of one year we apply the EOQ formula using expected demand for d.

The expected demand over one year is 12 x 6 = 72.

Then we have

$$EOQ = \sqrt{\{ \frac{2 \times 72 \times 150}{.20 \times 200} \}} = 23.24$$

As the EOQ has to be a whole number we recommend a value of 24.

The number of new orders placed per annum would then be 12 x 6/24 = 3.

If no buffer stock were set up then, with a lead time of 2 months, the reorder level would be equal to the expected demand over this period, namely 12 dishwashers.

If we use a buffer equal to b, then the reorder level is

R = 12 + b .

If D is the demand during lead time then this is Poisson with a mean of 12 and we require that

prob (D < 12 + b) = 0.05 .

35.3   (c)   (continued)

From the Poisson table (Table 2) it can be seen that

$$12 + b = 18$$

so that $\quad\quad b = 6$

The final recommendation, therefore, is that the retailer should use the following Reorder Level System:

A buffer stock of 6 dishwashers should be set up.

Reorders should be made when the total stock level reduces to 18 dishwashers.

An order for 24 dishwashers should be made when this occurs.

35.4   Explanatory Notes

1. Inventory control is a technique for determining an optimal or good policy for ordering and maintaining stock when there is a demand for the items being stocked and the retailer or manufacturer has control over an ordering policy.

2. There are basically two types of inventory control, namely a Reorder Level System and a Periodic Review System. In a reorder level system an order is placed for new stock when demand reduces the stock level to a certain level. Orders are placed, therefore, at variable time intervals. In a periodic review system the amount of stock is reviewed at equal time intervals and a decision is made whether to place an order or not. In this case orders are made at equal time intervals but the stock held, when an order is placed, is variable. In this example we are concerned with a reorder level system.

3. There are many costs involved in an inventory system and many of these can be associated with two principal costs, namely that for holding stock and that for placing an order for new stock. Stockholding costs are usually expressed as a percentage of the value of the stock held. In this example we regard an optimal policy as one that minimises the total cost, or the expected cost in the case of random demand etc. This is the usual criterion used although other criteria are possible; these are usually based on costs in some fashion or with the chances of running out of stock.

4. In a basic reorder level system there are two parameters to be chosen in order to define the system. The first is the amount of stock that needs to be ordered when an order is made. If this is chosen to minimise the cost then it is known as the Economic Order Quantity (EOQ). The second parameter is the Reorder Level and this is also chosen to minimise costs or to satisfy some other criterion. When the stock falls to the reorder level then an order is placed for a new amount of stock equal to the EOQ. This order may sometimes be received immediately but more often there is a time lag before it is received; this is known as the Lead Time.

5. These important quantities may be represented diagramatically and the following diagram represents a simple reorder level system when the rate of demand is constant.

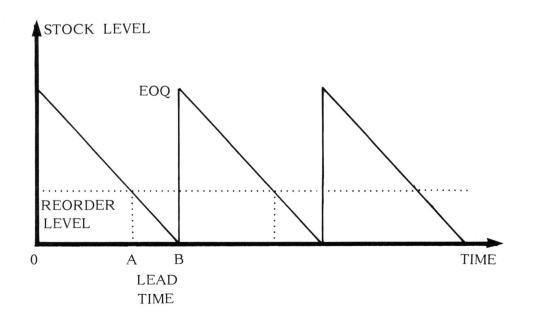

35.5   Explanatory Notes (continued)

5.   (continued)

The stock is reduced by demand until it reaches the reorder level; in the diagram this happens for the first time at time point A.   An amount of stock equal to the EOQ is ordered but this does not arrive until time point B due to the lead time.   In the meantime the stock level is further reduced by demand until it reaches zero at the end of the lead time at which point the ordered quantity of stock arrives and the process continues as before.   Note that we are able to allow stock to reduce to zero because the demand rate is constant and therefore predictable.   If the demand rate were not constant then this would be dangerous as it is likely that there would be unfulfilled demand.

6.   If we assume the simple reorder level system with constant demand and lead times as described above, then it is possible to minimise the total cost, due to stockholding and reordering, in order to determine the EOQ.   This is described in detail in the solution to the first part of the example and gives a simple 'square root' formula in terms of the two costs and the rate of demand.

7.   When demand is variable we simply replace the demand per unit time by the expected demand per unit time in the EOQ formula.   The reorder level is usually equal to td where t is the lead time and d is the demand rate; when lead time and demand are both variable t and d could be replaced by their expected values but, as we have stated previously, there would then be a good chance of a stockout.   To avoid this a Buffer Stock must be introduced; this is an extra amount of stock held to avoid a stockout.   With a buffer stock of b the reorder level then becomes td + b where t is the average lead time and d is the average demand per unit time.

8.   The choice of the buffer stock value, b, depends on the criterion to be satisfied.   In the example it has been chosen to ensure that the probability of a stockout is less than 5%.   Alternatively if the cost of a stockout were known then b could be chosen to minimise this cost; this would also require the determination of the probability of a stockout.

Main Points

•   A reorder level system is defined by the reorder level, the buffer stock and the economic order quantity.

•   There is a simple formula for the EOQ which may also be used when demand is variable.

•   The reorder level and buffer stock value may be chosen to give a small probability of stockout or to minimise costs if a cost can be associated with a stockout.

TABLE 1 : THE BINOMIAL DISTRIBUTION

This table shows a limited but typical tabulation of the Binomial Distribution appropriate to Example 5. Full values may be found in most standard statistical tables.

n = 5; p = 0.05

r	0	1	2	3
cumulative probabilities	.7738	.9774	.9988	1

n = 8; p = 0.3

r	0	1	2	3	4	5	6
cumulative probabilities	.0576	.2553	.5518	.8059	.9420	.9887	.9987

n = 10; p = 0.05

r	0	1	2	3	4	5
cumulative probabilities	.5987	.9139	.9885	.9990	.9999	1

TABLE 2 : THE POISSON DISTRIBUTION

This table shows a limited but typical tabulation of the Poisson Distribution. The values in the table are the upper cumulative probabilities. Full values may be found in most standard statistical tables.

m = 1.2

r	0	1	2	3	4	5	6	7
cumulative probabilities	1	.6988	.3374	.1205	.0338	.0077	.0015	.0003

m = 1.8

r	0	1	2	3	4	5	6	7
cumulative probabilities	1	.8347	.5372	.2694	.1087	.0364	.0104	.0026

m = 2.0

r	0	1	2	3	4	5	6	7
cumulative probabilities	1	.8647	.5940	.3233	.1429	.0527	.0166	.0045

TABLE 3 : THE STANDARD NORMAL DISTRIBUTION

Entries in the table below are the proportions of the area under the normal curve between the mean and **n** standard deviations.

n	.00	.01	.02	.03	.04	.05	.06	.07	.08	.09
0.0	.0000	.0040	.0080	.0120	.0160	.0199	.0239	.0279	.0319	.0359
0.1	.0398	.0438	.0478	.0517	.0557	.0596	.0636	.0675	.0714	.0753
0.2	.0793	.0832	.0871	.0910	.0948	.0987	.1026	.1064	.1103	.1141
0.3	.1179	.1217	.1255	.1293	.1331	.1368	.1406	.1443	.1480	.1517
0.4	.1554	.1591	.1628	.1664	.1700	.1736	.1772	.1808	.1844	.1879
0.5	.1915	.1950	.1985	.2019	.2054	.2088	.2123	.2157	.2190	.2224
0.6	.2257	.2291	.2324	.2357	.2389	.2422.	.2454	.2486	.2517	.2549
0.7	.2580	.2611	.2642	.2673	.2704	.2734	.2764	.2794	.2823	.2852
0.8	.2881	.2910	.2939	.2967	.2995	.3023	.3051	.3078	.3106	.3133
0.9	.3159	.3186	.3212	.3238	.3264	.3289	.3315	.3340	.3365	.3389
1.0	.3413	.3438	.3461	.3485	.3508	.3531	.3554	.3577	.3599	.3621
1.1	.3643	.3665	.3686	.3708	.3729	.3749	.3770	.3790	.3810	.3830
1.2	.3849	.3869	.3888	.3907	.3925	.3944	.3962	.3980	.3997	.4015
1.3	.4032	.4049	.4066	.4082	.4099	.4115	.4131	.4147	.4162	.4177
1.4	.4192	.4207	.4222	.4236	.4251	.4265	.4279	.4292	.4306	.4319
1.5	.4332	.4345	.4357	.4370	.4382	.4394	.4406	.4418	.4429	.4441
1.6	.4452	.4463	.4474	.4484	.4495	.4505	.4515	.4525	.4535	.4545
1.7	.4554	.4564	.4573	.4582	.4591	.4599	.4608	.4616	.4625	.4633
1.8	.4641	.4649	.4656	.4664	.4671	.4678	.4686	.4693	.4699	.4706
1.9	.4713	.4719	.4726	.4732	.4738	.4744	.4750	.4756	.4761	.4767
2.0	.4772	.4778	.4783	.4788	.4793	.4798	.4803	.4808	.4812	.4817
2.1	.4821	.4826	.4830	.4834	.4838	.4842	.4846	.4850	.4854	.4857
2.2	.4861	.4864	.4868	.4871	.4875	.4878	.4881	.4884	.4887	.4890
2.3	.4893	.4896	.4898	.4901	.4904	.4906	.4909	.4911	.4913	.4916
2.4	.4918	.4920	.4922	.4925	.4927	.4929	.4931	.4932	.4934	.4936
2.5	.4938	.4940	.4941	.4943	.4945	.4946	.4948	.4949	.4951	.4952
2.6	.4953	.4955	.4956	.4957	.4959	.4960	.4961	.4962	.4963	.4964
2.7	.4965	.4966	.4967	.4968	.4969	.4970	.4971	.4972	.4973	.4974
2.8	.4974	.4975	.4976	.4977	.4977	.4978	.4979	.4979	.4980	.4981
2.9	.4981	.4982	.4982	.4983	.4984	.4984	.4985	.4985	.4986	.4986
3.0	.4987	.4987	.4987	.4988	.4988	.4989	.4989	.4989	.4990	.4990

TABLE 4 : STUDENT'S t-DISTRIBUTION

p = probability tail   v = degrees of freedom

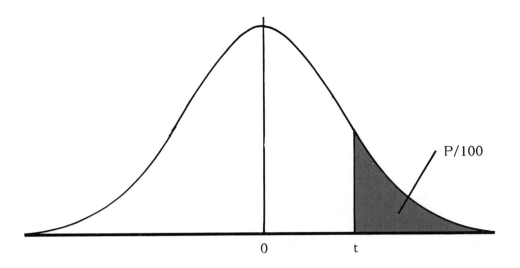

p	10	5	2.5	1	0.5	0.1	.005
v = 1	3.078	6.314	12.71	31.82	63.66	318.3	636.6
2	1.886	2.920	4.303	6.965	9.925	22.33	31.60
3	.638	.353	3.182	4.541	5.841	10.21	12.92
4	.533	.132	2.776	3.747	4.604	7.173	8.610
5	1.476	2.015	2.571	3.365	4.032	5.893	6.869
6	.440	1.943	.447	3.143	3.707	5.208	5.959
7	.415	.895	.365	2.998	.499	4.785	5.408
8	.397	.860	.306	.896	.355	.501	5.041
9	.383	.833	.262	.821	.250	.297	4.781
10	1.372	1.812	2.228	2.764	3.169	4.144	4.587
11	.363	.796	.201	.718	3.106	4.025	.437
12	.356	.782	.179	.681	3.055	3.930	.318
13	.350	.771	.160	.650	3.012	.852	.221
14	.345	.761	.145	.624	2.977	.787	.140
15	1.341	1.753	2.131	2.602	2.947	3.733	4.073
16	.337	.746	.120	.583	.921	.686	4.015
17	.333	.740	.110	.567	.898	.646	3.965
18	.330	.734	.101	.552	.878	.610	.922
19	.328	.729	.093	.539	.861	.579	.883
20	1.325	1.725	2.086	2.528	2.845	3.552	3.850
21	.323	.721	.080	.518	.831	.527	.819
22	.321	.717	.074	.508	.819	.505	.792
23	.319	.714	.069	.500	.807	.485	.768
24	.318	.711	.064	.492	.797	.467	.745

TABLE 5 : PERCENTAGE POINTS OF THE χ^2-DISTRIBUTION

p = probability tail   v = degrees of freedom

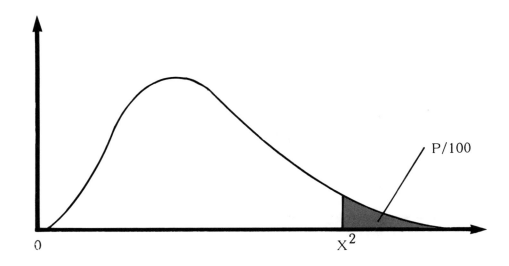

p	10	5	2.5	1	0.5	0.1	0.05
v = 1	2.706	3.841	5.024	6.635	7.879	10.83	12.12
2	4.605	5.991	7.378	9.210	10.60	13.82	15.20
3	6.251	7.815	9.348	11.34	12.84	16.27	17.73
4	7.779	9.488	11.14	13.28	14.86	18.47	20.00
5	9.236	11.07	12.83	15.09	16.75	20.52	22.11
6	10.64	12.59	14.45	16.81	18.55	22.46	24.10
7	12.02	14.07	16.01	18.48	20.28	24.32	26.02
8	13.36	15.51	17.53	20.09	21.95	26.12	27.87
9	14.68	16.92	19.02	21.67	23.59	27.88	29.67
10	15.99	18.31	20.48	23.21	25.19	29.59	31.42
11	17.28	19.68	21.92	24.72	26.76	31.26	33.14
12	18.55	21.03	23.34	26.22	28.30	32.91	34.82
13	19.81	22.36	24.74	27.69	29.82	34.53	36.48
14	21.06	23.68	26.12	29.14	31.32	36.12	38.11
15	22.31	25.00	27.49	30.58	32.80	37.70	39.72
16	23.54	26.30	28.85	32.00	34.27	39.25	41.31
17	24.77	27.59	30.19	33.41	35.72	40.79	42.88
18	25.99	28.87	31.53	34.81	37.16	42.31	44.43
19	27.20	30.14	32.85	36.19	38.58	43.82	45.97
20	28.41	31.41	34.17	37.57	40.00	45.31	47.50
21	29.62	32.67	35.48	38.93	41.40	46.80	49.01
22	30.81	33.92	36.78	40.29	42.80	48.27	50.51
23	32.01	35.17	38.08	41.64	44.18	49.73	52.00
24	33.20	36.42	39.36	42.98	45.56	51.18	53.48

TABLE 6 : CONFIDENCE INTERVALS

Situation	Formula	Sampling Distribution
For a mean (large sample)	$\bar{x} \pm z\sigma/\sqrt{n}$	z is standard normal
For a mean (small sample)	$\bar{x} \pm ts/\sqrt{n}$	t is Student's t on $(n-1)$ d.f.
Difference in means (large sample)	$(\bar{x} - \bar{y}) \pm z\sqrt{\{\sigma_1^2/n_1 + \sigma_2^2/n_2\}}$ σ_1 is S.D. of x σ_2 is S.D. of y	z is standard normal
Proportion	$r/n \pm z\sqrt{\{r(n-r)/n^3\}}$ (r successes in n trials)	z is standard normal
Difference in Proportions	$(r_1/n_1 - r_2/n_2) \pm z\sqrt{\{r_1(n_1-r_1)/n_1^3 + r_2(n_2-r_2)/n_2^3\}}$ r_1 successes in n_1 trials r_2 successes in n_2 trials	z is standard normal

TABLE 7 : SIGNIFICANCE TESTS

Situation	Test Statistic	Sampling Distribution
For a mean, μ (large sample)	$(\bar{x} - \mu)\sqrt{n}/\sigma$	Standard normal
For a mean, μ (small sample)	$(\bar{x} - \mu)\sqrt{n}/S$	Student's t on $(n-1)$ d.f.
Equality of means (large sample)	$(\bar{x} - \bar{y})/\sqrt{\{\sigma_1^2/n_1 + \sigma_2^2/n_2\}}$ σ_1 is S.D. of x σ_2 is S.D. of y	Standard normal
Proportion, p	$(r/n - p)/\sqrt{\{p(1 - p)/n\}}$	Standard normal
Difference in Proportions	$(r_1/n_1 - r_2/n_2)/\sqrt{\{a(1-a)(1/n_1 + 1/n_2)\}}$ where $a = (r_1+r_2)/(n_1+n_2)$ (r_1 successes in n_1 trials) (r_2 successes in n_2 trials)	Standard normal

INDEX

INDEX